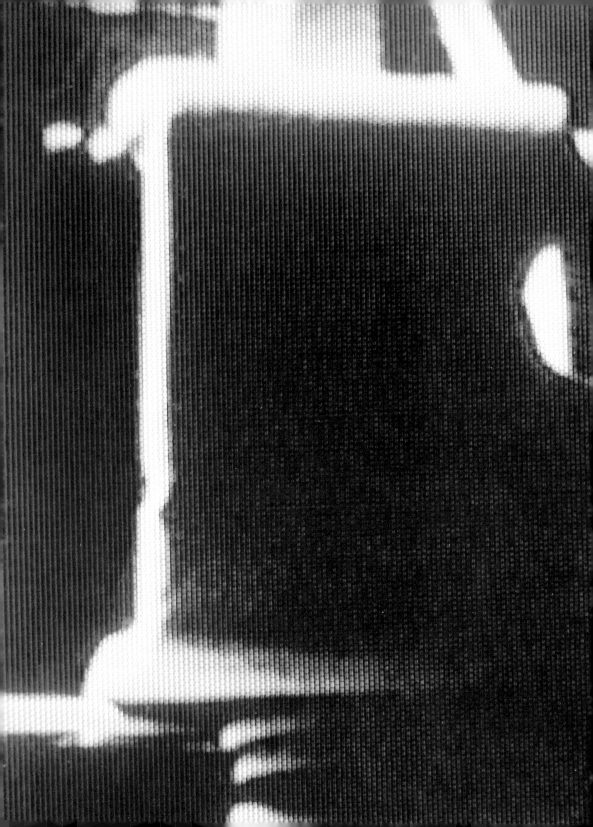

建筑与废物

王昀

中国电力出版社
CHINA ELECTRIC POWER PRESS

图书在版编目（CIP）数据

跨界设计.建筑与废物/王昀著.--北京：中国电力出版社，2016.4
ISBN 978-7-5123-9076-8

Ⅰ.①跨… Ⅱ.①王… Ⅲ.①建筑设计 Ⅳ.①TU2

中国版本图书馆CIP数据核字(2016)第050876号

感谢北京建筑大学建筑设计艺术研究中心
建设项目的支持

中国电力出版社出版发行
北京市东城区北京站西街19号 100005
http://www.cepp.sgcc.com.cn
责任编辑：王 倩
封面设计：方体空间工作室（Atelier Fronti）
版式设计：张捍平
责任印制：蔺义舟
责任校对：太兴华
英文翻译：陈伟航
北京盛通印刷股份有限公司印制·各地新华书店经售
2016年4月第1版·第1次印刷
787mm×1092mm 1/16·17印张·320千字
印数：1-2000册
定价：58.00元

敬告读者
本书封底贴有防伪标签，刮开涂层可查询真伪
本书如有印装质量问题，我社发行部负责退换
版权专有 翻印必究

内容提要

建筑是时代表征物，其身上凝聚有诸多时代的信息和相应时代技术、艺术所赋予的烙印。由于建筑自身在时代发展过程中的相对滞后性，当一种科技的结晶物成为时代特征的瞬间，其实作为科技产品的本身已经是一个落后的产品了，本书将这些似乎已经过时并是以"垃圾"和"废品"的面貌来呈现的物品再次重生，将其作为建筑的摹本，并将一组组全新的建筑的视觉图景加以呈现是本书的立论，也是本书所提示的一种全新的思考与设计方法论。本书适合建筑、艺术、设计相关专业师生和建筑师、学术研究者阅读。

Abstract

Architecture, as representation of times, is not only condensed with huge time-related information but also marked by respective technology and art. Due to relative lagging nature of architecture compared to development of times, once a technology crystallizes into a feature of time, the technology product itself actually already becomes an obsolete product. This book intends to regenerate these objects, which seem to be obsoleted and appeared in the form of "rubbish" and "junk", use them as the prototype of architecture and the display sets of brand-new architectural visual scenes. This book, proposing a brand-new way of thinking and designing methodology, is suitable for teachers and students majoring in architecture, art, design and relevant majors as well as architects, researchers.

Architecture and Junk
WangYun

序 Preface

　　《建筑与废物》这本书中所涉及的一系列我所设计的未建成的作品，是笔者于20世纪90年代中期所开始进行的一系列关于未来建筑思考的展示。起因是伴随我学习外语的小录音机出现了卡带的故障，我试图对其修理的初衷却成为将其变成了废品的行为过程。然而当对这个"废品"进行"解剖"的瞬间，猛然意识到眼前这个"废品"与建筑之间的关联。建筑是时代的产物，建筑反映相应时代判断的本身，表明建筑在进行构思和搭建的过程中眼前已经摆放着时代的"摹本"或"样本"，技术是一个时代的表征，由技术所产生的一系列的结晶体也会瞬间地由于技术的再发展而成为之前技术条件下的废品和垃圾。而所谓建筑反映相应时代的命题也就只能是面对已经完成的结晶体。而不幸的是当你面对这些技术的结晶体时，这些结晶体已经被真正的技术所超越，而建筑所反映的也不过是相应时代的遗弃品。我们这个时代的周边充满了多种结晶体，可这些结晶体对我们来讲还仍然没有企及，或许它们已经成为了时代技术的"垃圾"，但是如同那个曾经给我带来便利的小录音机那样，它也曾经是一个新技术的结晶体。但它在新技术的发展过程下，的确地也已经成为了"废品"和"垃圾"，但是这个"垃圾"却可以转而成为建筑的"摹本"。正是在这样的一种思考的前提下，在我们还在关注着或者不断地将几千年以前的技术的"垃圾"作为我们今天"摹本"的时代，能否让我们关注一下离我们最近的这些"垃圾"所带给我们的新的感受、新的体验和新的视觉是这本书的终极意图，由于每一个过往的结晶体一定曾凝聚过相应时代的尖端技术，从这个意义上说今天的"垃圾"就是明天的传统。

　　给本书起名叫《建筑与废物》完全是由于一个纠结的过程。"建筑与垃圾"这个名字可能更符合这本书的性格，然而建筑和垃圾的提法引来了诸多的理解上的歧义。而最终采用了现在这本书的命名。我想重申的一点是：我所说的"垃圾"的概念，并不是一个日常生活中所理解的垃圾的意思，是指那些被时代扔掉的并可以从中寻找到视觉含义的"结晶体"要素。在我看来：看似没有用的东西，甚至垃圾，都有可能直接地指向建筑。

The book *Architecture and Junk* touches on a series of unestablished architecture works I designed, which demonstrates a series of contemplation on future architecture I started at the first half of the 1990s. The origin is the action process as I tried in vain to repair a malfunctioning tape recorder that accompanied my foreign language study but only turned it into a damaged junk. But when I was "disassembling" this "junk", I suddenly realized the relation between architecture and this "junk" at hand. Architecture, as product of times, reflects judgment of the respective time itself and indicates that the "prototype" or "sample" of the time is already before one's eyes when designing and constructing the architecture. As technology is the representation of a certain time, a series of crystallized objects resulted from technology will instantaneously turn into junk and rubbish of the previous technology due to further technology development. The statement that architecture reflects respective time is only applicable on crystallized objects already completed. Unfortunately when you are facing such crystallized objects of technology, they are already transcended by real technology, leaving architecture only to reflect derelict of respective time. There are various crystallized objects around us in this age, but they are not within our reach yet, or they have already become "rubbish" of respective technology. However, similar to that small tape recorder that once brought convenience to me, it was once a crystallized object of a new technology then. During the development process of new technology, it indeed became "junk" and "rubbish", yet this "rubbish" can be transformed into "prototype" of architecture. With precondition of such a thinking, while we still pay attention to or keep transforming technology "rubbish" thousands of years ago into our "prototype" today, it is the ultimate intention of this book for us to focus on new feelings, new experience and new vision brought by these "rubbish" most close to us. As each past crystallized object definitely once concentrated cutting-edge technology of respective ages, from this sense, "rubbish" of today is tradition of tomorrow.

To give this book the name of *Architecture and Junk* is purely due to an entangled procedure. The name of "Architecture and Rubbish" may better comply with the nature of this book, but the wording of architecture and rubbish results in many ambiguous meanings in understanding. Therefore this book is named as it is now. I want to reiterate that my concept of "rubbish", unlike the rubbish commonly we understood in daily life, refers to elements of "crystallized objects" discarded by ages and marked with visual meanings that can be found from within. In my opinion, those objects seemingly of no use, or even rubbish, may directly point to architecture.

王昀
WangYun
2016年01月

目录

序

导读 3

从科技结晶体、废物和垃圾中发现与建筑相关的18个未建成案例 5

1	现代美术馆	8
2	当代博物馆	25
3	新世纪传媒中心	35
4	科技大厦	45
5	CBD 国际城	54
6	都会广场	75
7	大剧院	89
8	图书馆	105
9	光宅	143
10	文化宫	159
11	大木仓密集城市	171
12	希望小学	183
13	大观园规划	189
14	数学家住宅	195
15	贸易市场	204
16	公园住宅	215
17	摄影家沙龙	223
18	未来城	231

"废物"观察术 246

作者简介 255

Contents

Preface

Introduction   3

18 unestablished cases related to architecture, discovered from   5
crystallized objects of science and technology, junk and rubbish

1	Modern Art Gallery	8
2	Contemporary Museum	25
3	New Century Media Center	35
4	Science Mansion	45
5	CBD International City	54
6	Metropolis Plaza	75
7	Grand Theatre	89
8	The Library	105
9	House of Light	143
10	Cultural Palace	159
11	Damucang Intensive City	171
12	Hope Primary School	183
13	Grand View Garden Plan	189
14	Mathematicians Residence	195
15	Trade Market	204
16	Park Residence	215
17	Photographers Saloon	223
18	Future City	231

How to observe "junk"   246

About the author   255

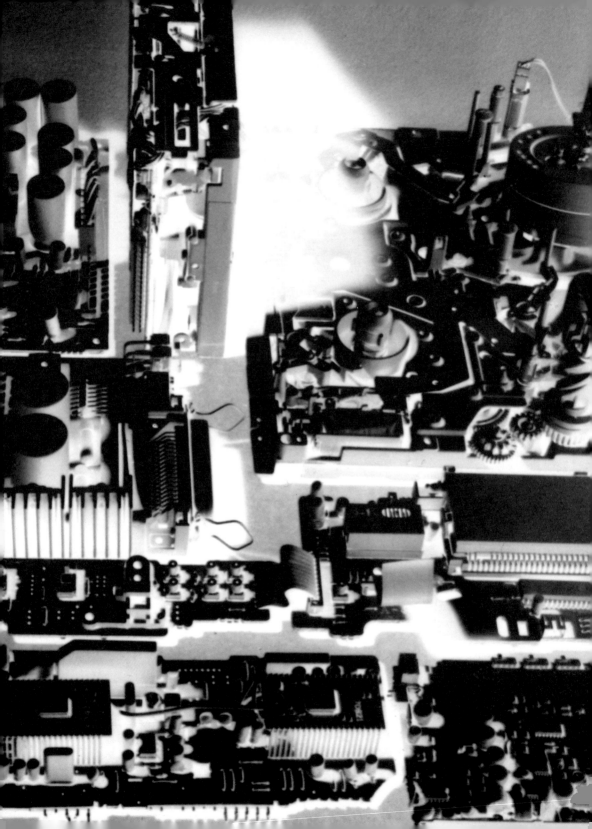

看似没有用的东西，有可能称为建筑
Those seemingly useless objects may be called architecture

左图：这不是一组废物
Left figure: these are not junk

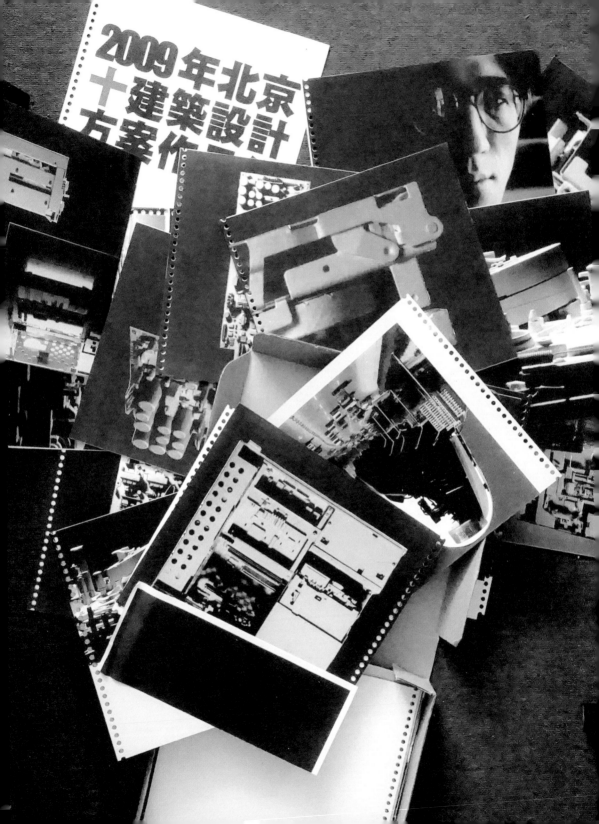

导读 Introduction

　　本质上讲，这本小册子中所呈现的每一个"场景"均是以"置换"作为概念而引发的一系列与建筑关联的设计展示，以自由的眼睛去审视和关注当代生活周边的"场景""结晶体"（甚至"垃圾"），能够获得自由与开放的未来图景。书中所罗列的18个未建成的相关案例，均为个人自1995年始，对自己"修坏"的物品进行肢解观察时萌发的设计结果，有时为了能尽早地看到某种景象，更期待新买回来的新科技的结晶体能够尽快坏掉，如此这样"变态"的逻辑一直持续到这本书的编辑结束。为了本书的出版，家里的一台尚未用坏的电视机在它完成了最后的"任务"的瞬间成为了本书的"未来城"。书中每一个场景的背后都有一段由"新"到"旧"再到"置换"和"重生"的过程："国家大剧院"的未建成项目是1996年用坏掉的一台录像机重生后所得到的结果；"现代美术馆"是上面所谈到的1995年被自己修坏的录音机……

　　1999年从当时积攒的十几个重生的结晶体中，选出了10个希望能够在2009年在国内完成真正的设计，并冠以"2009年北京十大建筑设计方案"。而今，距离完成这些方案已经过去了6年，其中些许"意向"也因"间断"有所变迁，书中的18个相关设计是在之前所思考的基础上重新增加项目所进行的整体呈现。

In essence, every "scene" appeared in this book uses "transformation" as a concept to trigger a series of design display related to architecture. Using eyes of freedom to examine and notice surrounding "scenes" and "crystallized objects" (even "rubbish"), one can obtain future vision of freedom and openness. The 18 related cases of unestablished architecture projects listed in this book are all design results when I disassembled and observed objects damaged by myself in repair since 1995. Sometimes I even expected newly purchased crystallized objects of new technology may break down as soon as possible so that I could quickly see a certain scene. Such an "abnormal" logic stayed on until the completion of editing this book. For the purpose of publishing this book, a television set still functioning at home, upon completing its last "mission", instantaneously became "Future City" in this book. Behind every scene in this book, there is a procedure from "new" to "old" and then to "transformation" and "regeneration": the unestablished project of "National Grand Theatre" is the regeneration result of a VCR broken in 1996, while "Modern Art Gallery" is the tape recorder damaged by myself during repair in 1995 mentioned above.

In 1999, from over a dozen regenerated crystallized objects I accumulated by then, I picked out 10 objects that I hoped to finish real design in China by 2009 and entitled them as "2009 Ten Great Architecture Design Plans in Beijing". Now six years elapsed since I completed these plans, and some "intentions" change due to "interruptions". The 18 related designs are overall representation based on previous thinking with newly added items.

左图：1998年制作的《2009年北京十建筑设计方案作品集》
Left figure: 2009 Ten Great Architecture Design Plans in Beijing, made in 1998

从科技结晶体、废物和垃圾中发现与建筑相关的18个未建成案例
18 unestablished cases related to architecture, discovered from crystallized objects of science and technology, junk and rubbish

以下将以视觉呈现的方式逐一列出未建成的18个案例：现代美术馆、当代博物馆、新世纪传媒中心、科技大厦、CBD国际城、都会广场、大剧院、图书馆、光宅、文化宫、大木仓密集城市、希望小学、大观园规划、数学家住宅、贸易市场、公园住宅、摄影家沙龙、未来城。

The following 18 cases of unestablished projects are listed in a visualized way one by one: Modern Art Gallery, Contemporary Museum, New Century Media Center, Science Mansion, CBD International City, City Plaza, Grand Theatre, The Library, House of Light, Cultural Palace, Damucang Intensive City, Hope Primary School, Grand View Garden Plan, Mathematicians Residence, Trade Market, Park Residence, Photographers Saloon, Future City.

1

现代美术馆
Modern Art Gallery

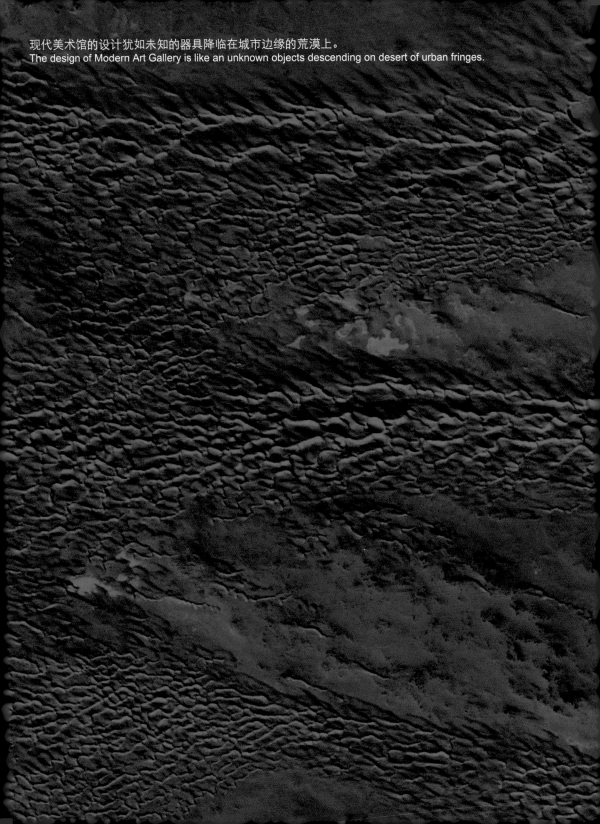

现代美术馆的设计犹如未知的器具降临在城市边缘的荒漠上。
The design of Modern Art Gallery is like an unknown objects descending on desert of urban fringes.

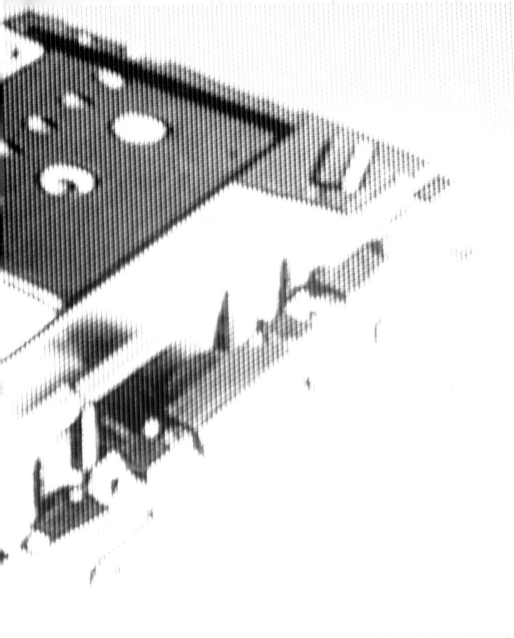

现代美术馆是以收藏、研究、展示现代艺术家作品的艺术博物馆。现代美术馆收藏美术作品3万余件。美术馆举行国内外美术学术交流，建立现代美术史料、艺术档案，编辑出版藏品画集、理论文集等。
Modern Art Gallery is an art museum that collects, studies and exhibits works of modern artists. With over 30,000 art works in collection, it holds domestic and international art academic exchanges, establishes modern art records and archives, edits and publishes painting collections and theses.

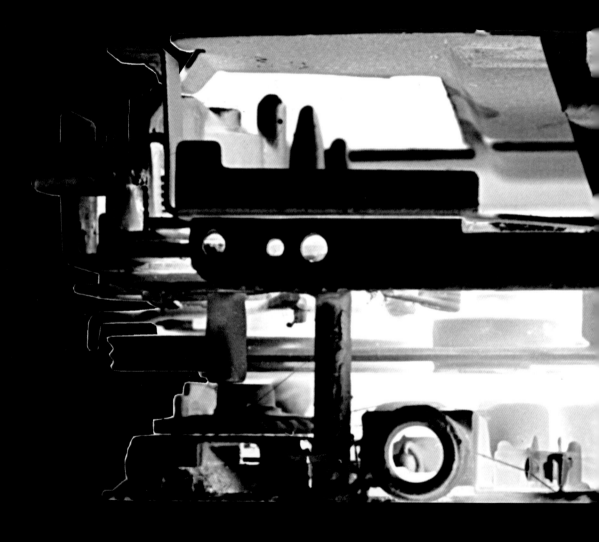

现代美术馆东立面
East facade of Modern Art Gallery

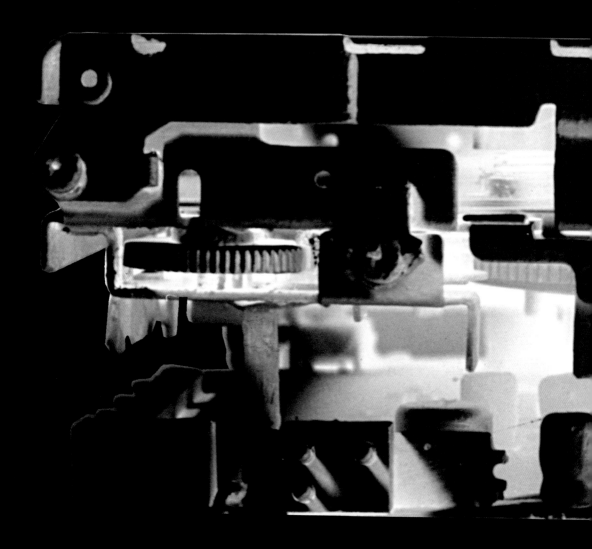

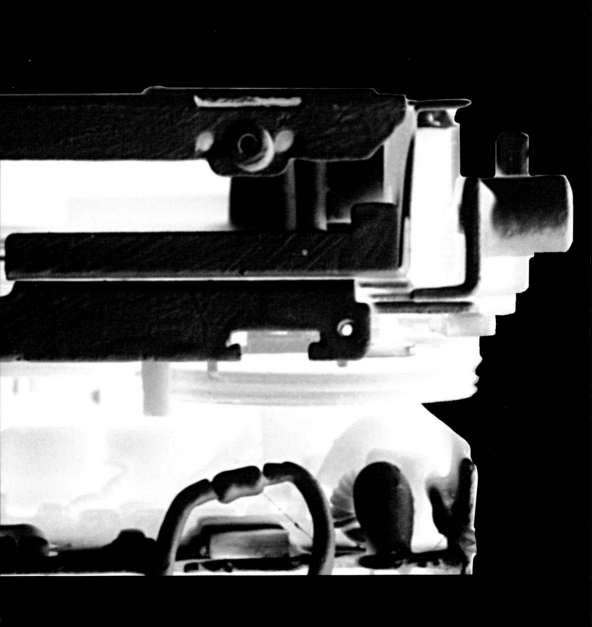

现代美术馆西立面
West facade of Modern Art Gallery

现代美术馆南立面
South facade of Modern Art Gallery

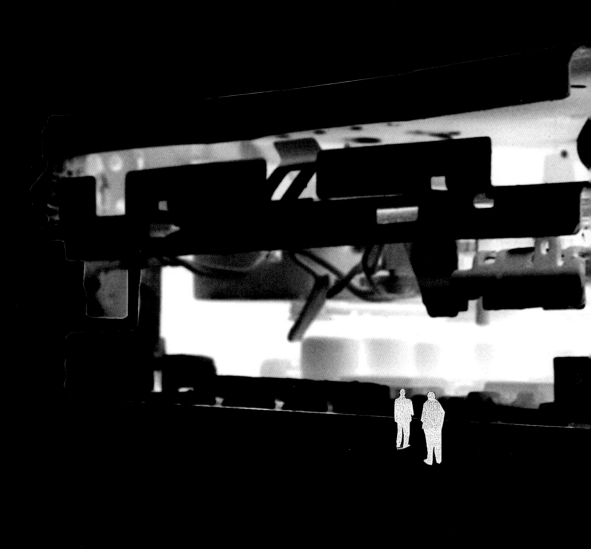

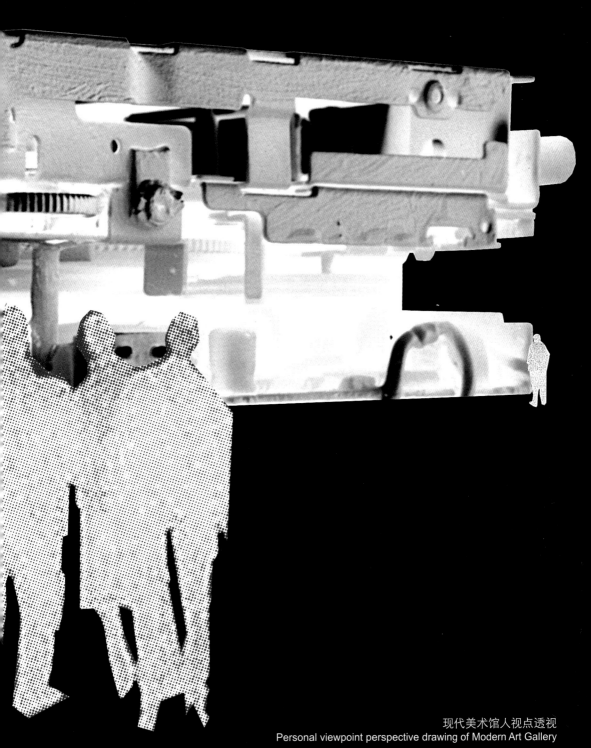

现代美术馆人视点透视
Personal viewpoint perspective drawing of Modern Art Gallery

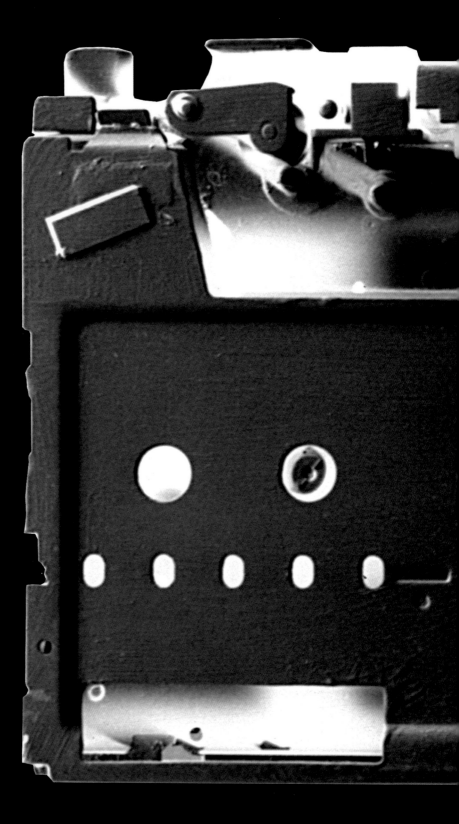

屋顶平面
Roof plan

2

当代博物馆
Contemporary Museum

当代博物馆鸟瞰
Aerial view of Contemporary Museum

当代博物馆
Contemporary Museum

当代博物馆位于北京玉渊潭公园内,建筑建在公园水面之上,内设展厅、书店、图书资料室、小讲堂。

Located in Beijing Yuyuantan Park, Contemporary Museum is built above water surface and equipped with exhibition room, book store, library and lecture hall.

当代博物馆鸟瞰
Aerial view of Contemporary Museum

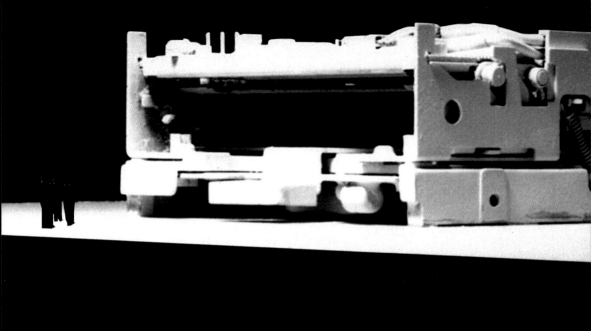

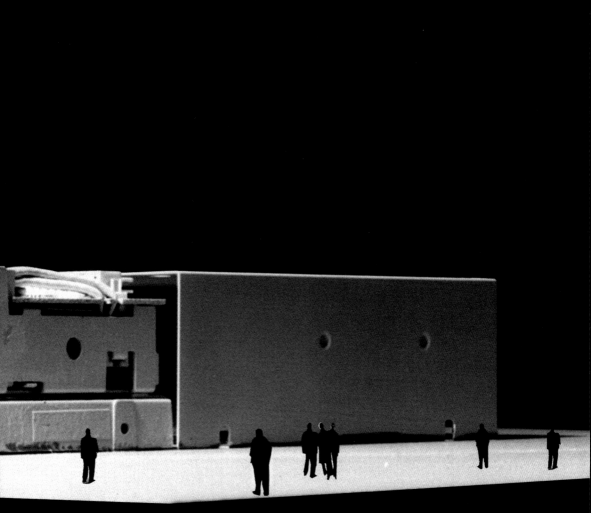

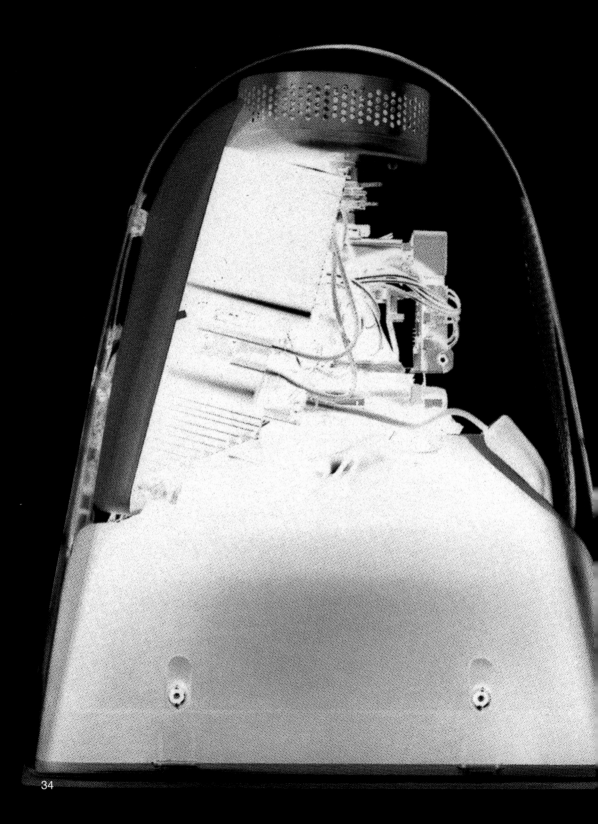

3

新世纪传媒中心
New Century Media Center

新世纪传媒中心
New Century Media Center

新世纪传媒中心是以信息收集、编辑、发布和监管为主要功能的综合处理平台,巨大的体量位于城市的中心区域,成为城市媒体和声音的象征。

New Century Media Center is a comprehensive processing platform with major functions of collecting, editing, releasing and monitoring information. With bulky size and central location, it is the symbol of urban media and voice.

雾霾中的新世纪媒体中心
New Century Media Center in smog

位于城市中心的新世纪媒体中心
New Century Media Center in downtown

科技大厦
Science Mansion

科技大厦
Science Mansion

西单科技大厦,位于西单路口东北角,原为科技画廊所处的位置。科技大厦是以科技为主题,含有展览、教育、培训、酒店功能的城市综合体。大厦与西单文化广场及西单地区地下广场相互连通,地铁可直接通达这里。

Xidan Science Mansion is located at the northeast corner of Xidan road junction formerly occupied by Science Gallery. With the theme of science, this is an urban complexity with functions of exhibition, education, training and hotel. The mansion interconnects with Xidan Culture Plaza and Xidan Underground Square, directly accessible by subway.

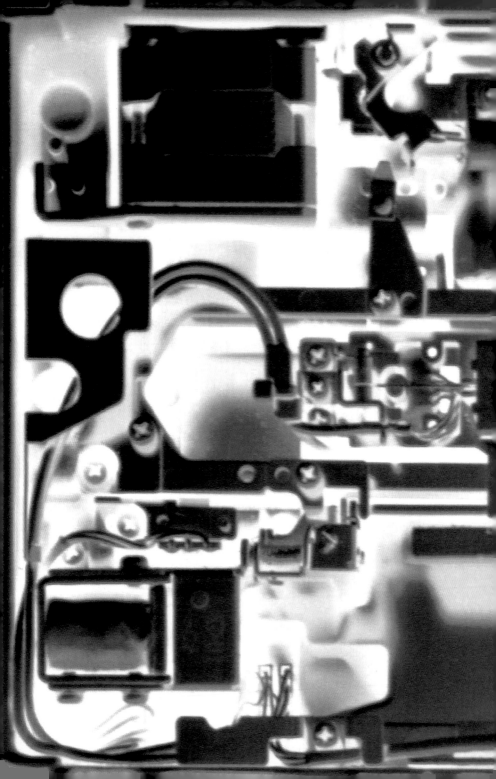

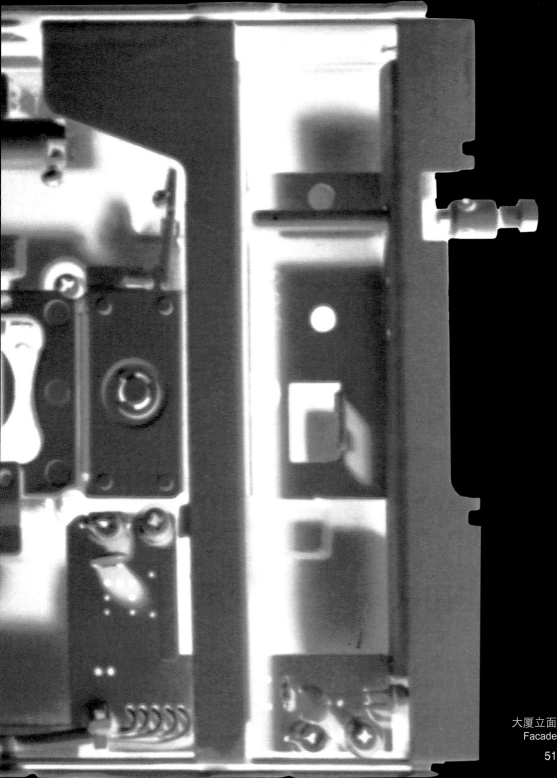

大厦立面
Facade

从西单路口西南角远望科技大厦
Distant view of Science Mansion from southwest corner of Xidan road junction

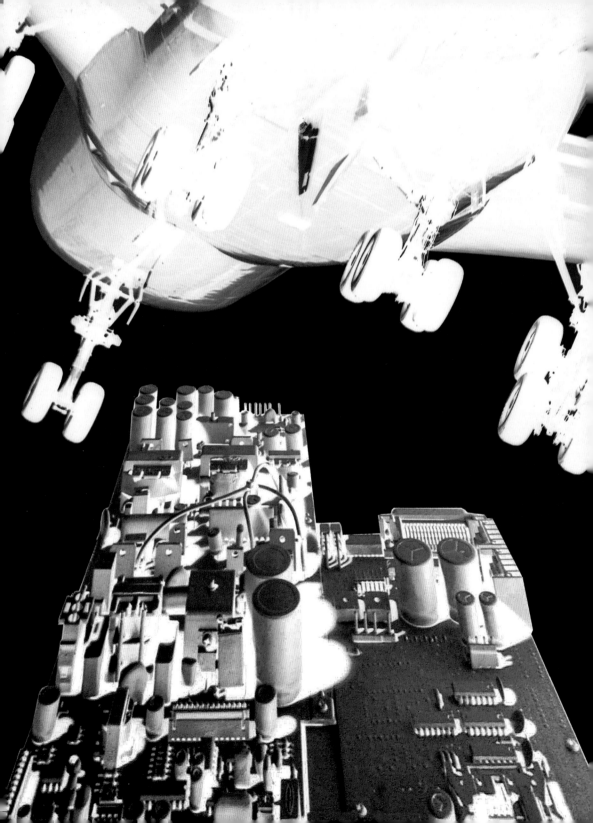

CBD国际城
CBD International City

国际城位于北京国贸东北角,是一个聚集了办公、会议、展示、交流、餐饮、商业等多项功能的综合区。
CBD International City, located at the northeast corner of Beijing World Trade Center, is a complexity with multiple functions such as office, conference, exhibition, communication, catering and shopping.

CBD国际城鸟瞰
Aerial view of CBD International City

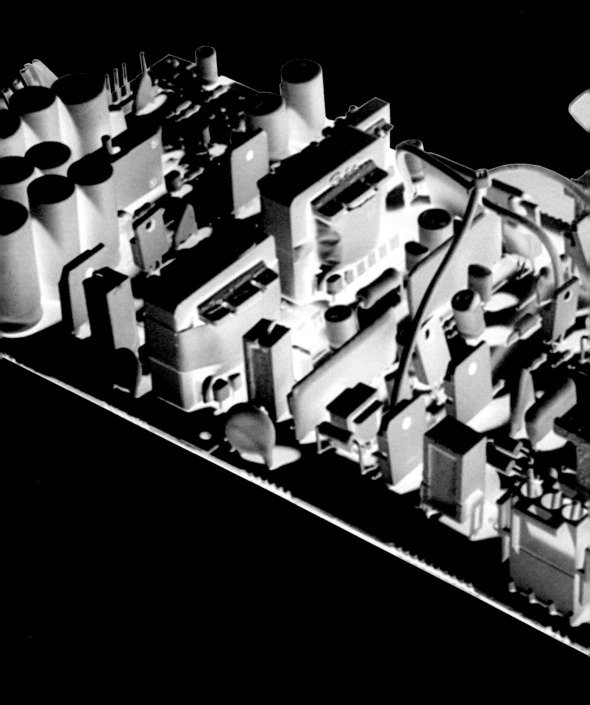

CBD国际城鸟瞰
Aerial view of CBD International City

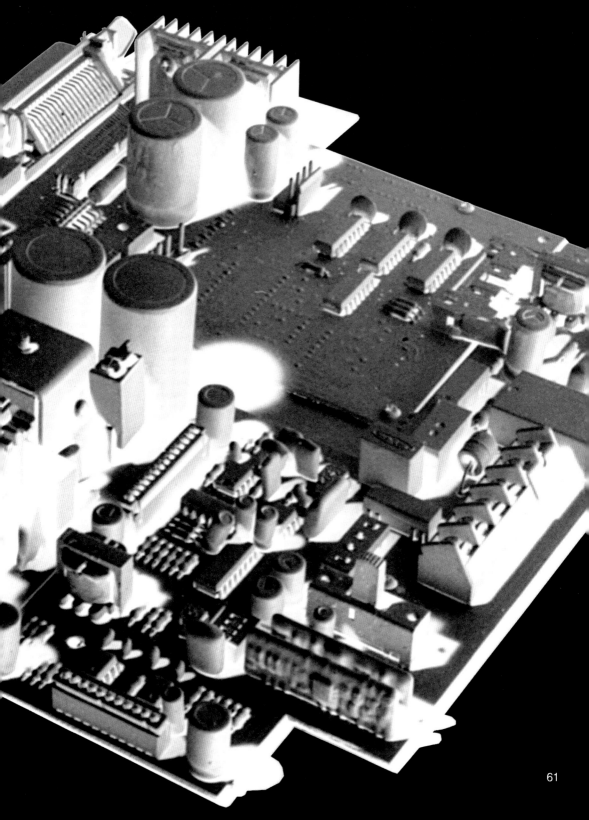

CBD国际城透视
Perspective view of CBD International City

CBD国际城街景
Street view of CBD International City

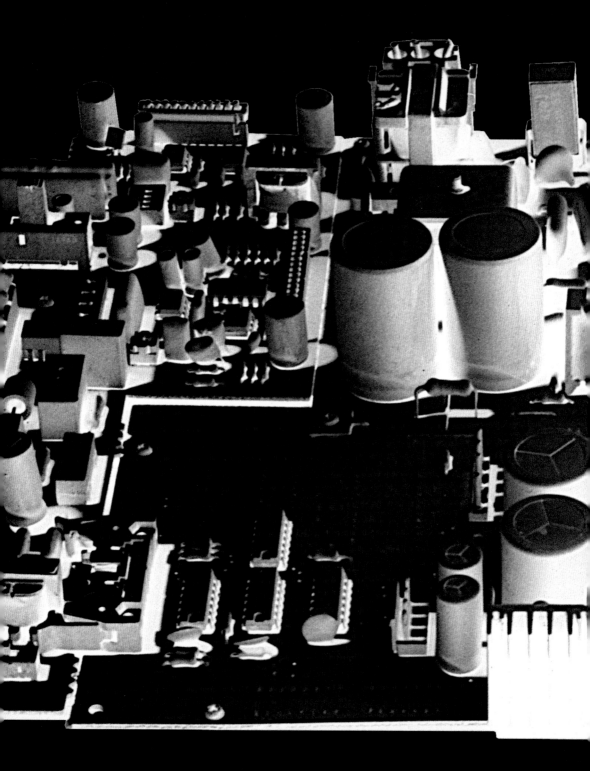

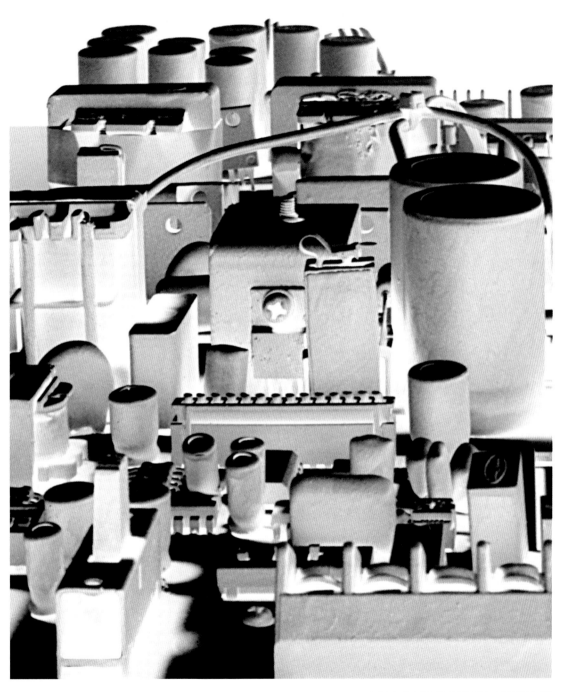

左图和右图均为CBD国际城近景鸟瞰
Both left and right figures are close aerial view of CBD International City

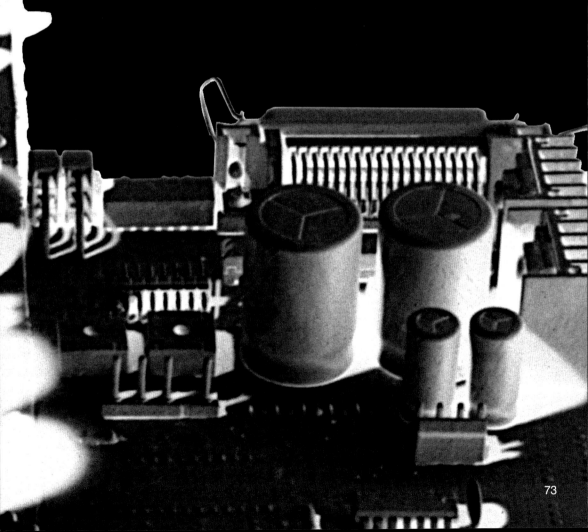

6
都会广场
Metropolis Plaza

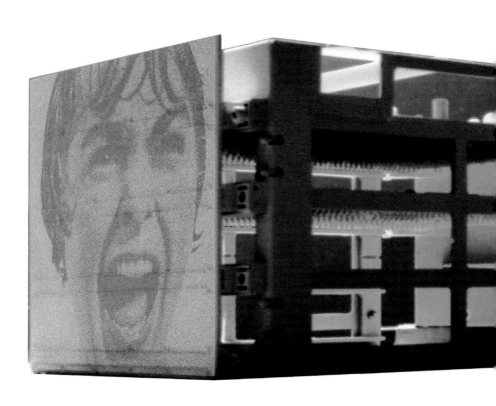

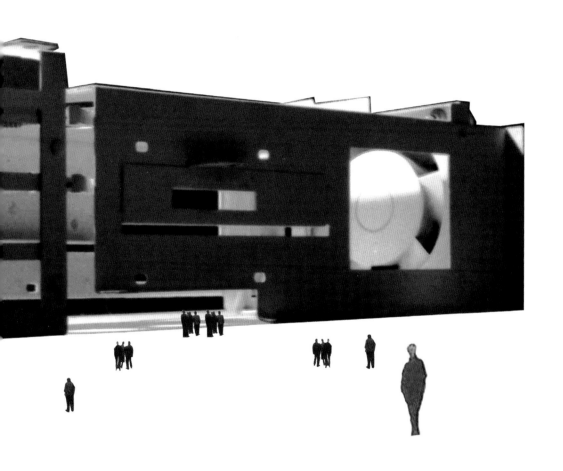

都会广场透视
Perspective view of Metropolis plaza

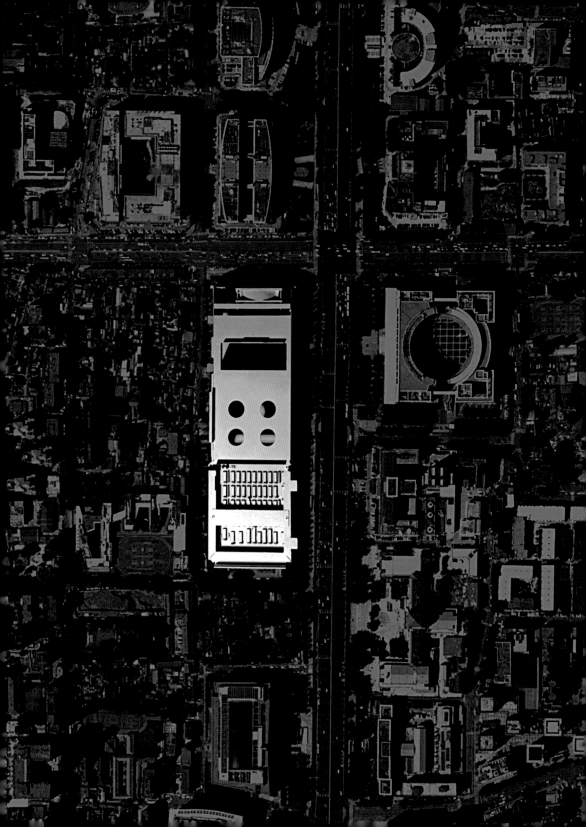

都会广场
Metropolis plaza

都会广场位于西长安街复兴门内大街南侧，紧邻闹市口大街路口。广场作为未来都市中心的生活中心，内设百货商店、办公、酒店、剧场、饮食街、多功能大厅、观景台、停车场等，是一个大型的综合型建筑。

Metropolis plaza is on the south side of Fu Xing Men Nei Street of West Chang'an Avenue, close to Nao Shi Kou Street. As the life center of future urban center, the plaza is a large complexity with functions of department store, office, hotel, theatre, catering street, multifunctional hall, observation deck, parking ground and so on.

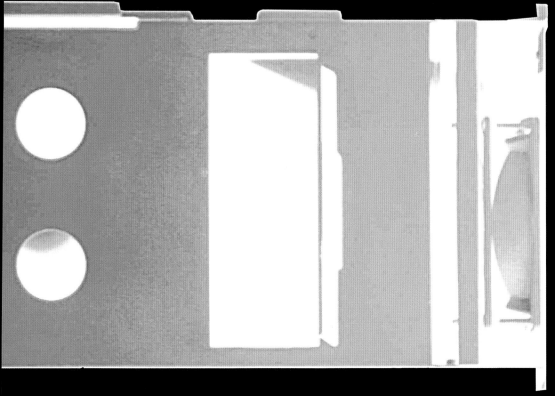

都会广场屋顶平面
Roof of Metropolis plaza

在长安街北侧向南望都会广场
Southward view of Capital Plaza from north side of Chang'an Avenue

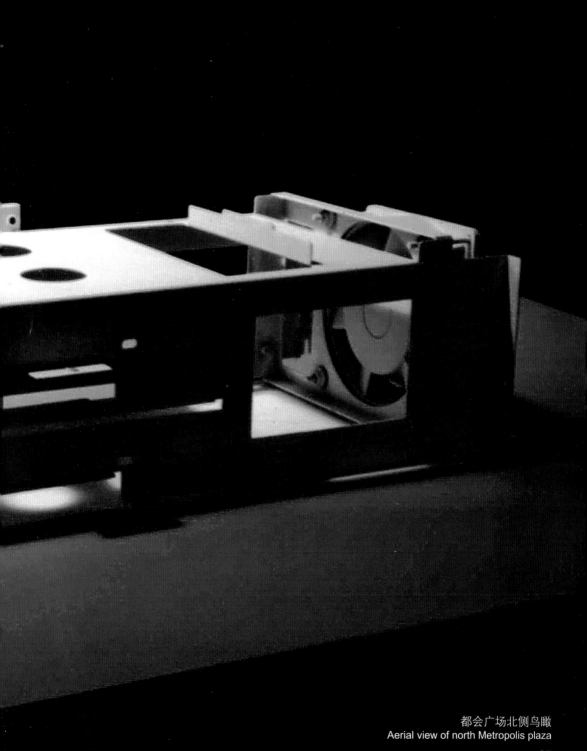

都会广场北侧鸟瞰
Aerial view of north Metropolis plaza

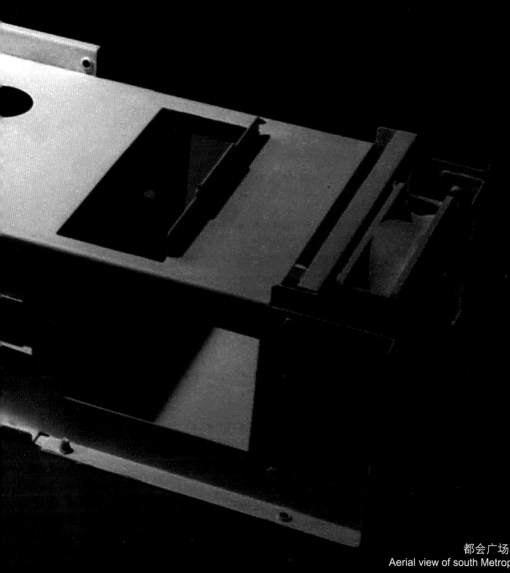

都会广场南侧鸟瞰
Aerial view of south Metropolis plaza

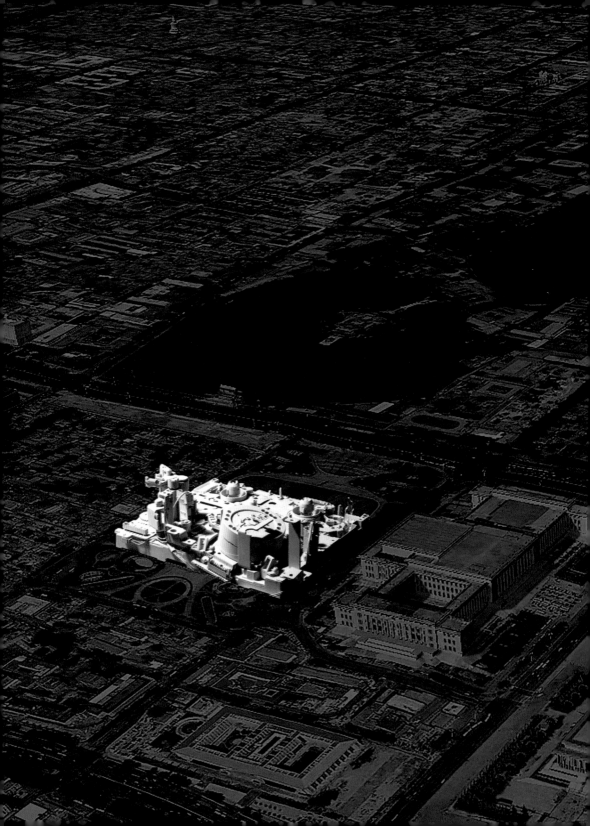

7

大剧院
Grand Theatre

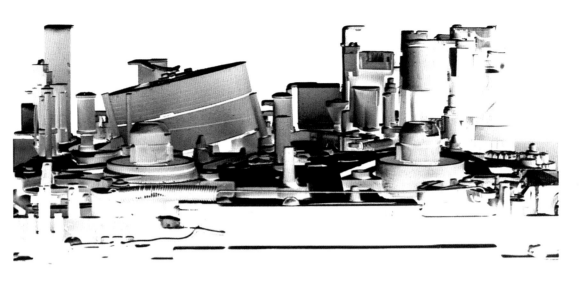

本大剧院的总占地面积11.89万平方米，总建筑面积约16.5万平方米，其中主体建筑10.5万平方米，地下附属设施6万平方米。大剧院的构思原初于录像机的机芯造型，作为影像收放功能技术的结晶体，将其形态作为大剧院的整体意向形态恰如其分。

Grand Theatre occupies a total land area of 118,900 square meters, with total floor area of 165,000 square meters (105,000 square meters of main building and 60,000 square meters of affiliated underground facilities). The conception of the Grand Theatre originated from mechanism of VCR. As a crystallized object for video recording and playing technology, it is very suitable to use this pattern to reflect overall pattern of the Grand Theatre.

大剧院鸟瞰
Aerial view of Grand Theatre

大剧院入口人视点透视
Personal viewpoint perspective view of Grand Theatre entrance

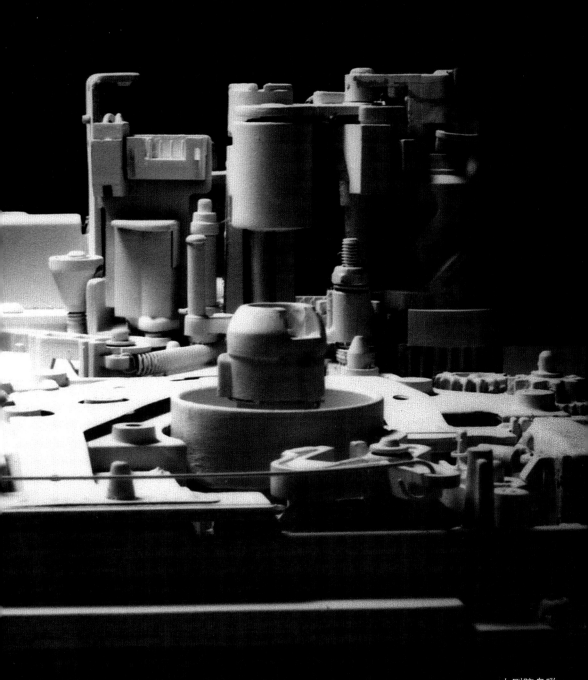

大剧院鸟瞰
Aerial view of Grand Theatre

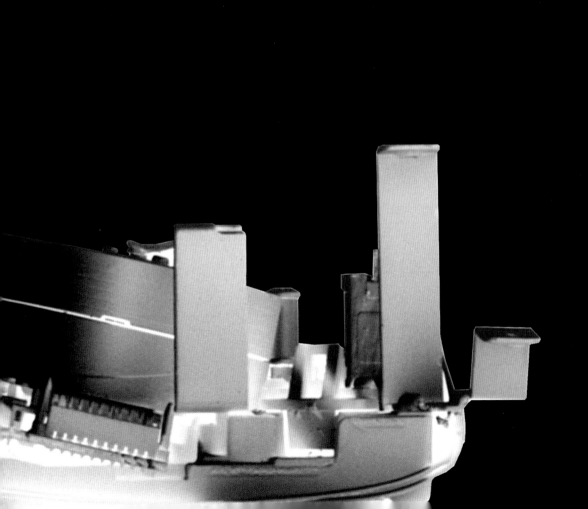

大剧院剧场小剧场空间
Small theatre space of Grand Theatre

大剧院剧场局部透视
Partial perspective of the theatre part of Grand Theatre

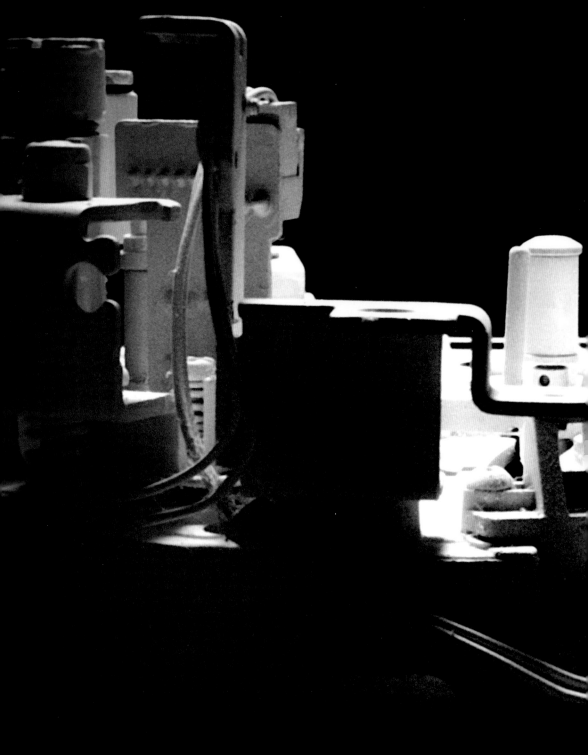

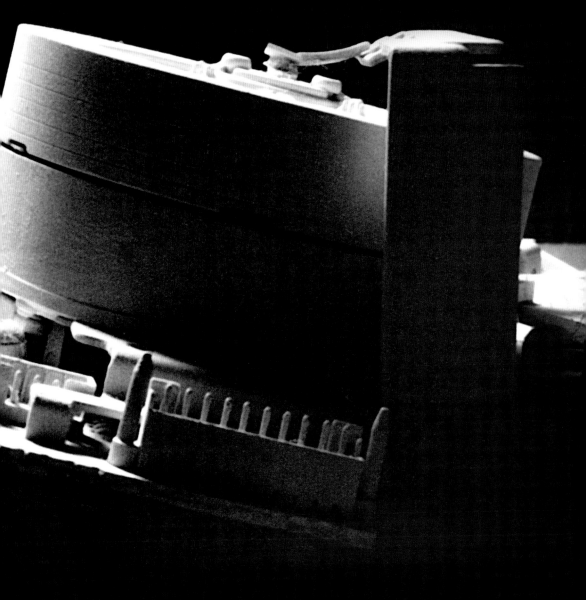

大剧院局部透视
Partial perspective of Grand Theatre

图书馆
The Library

图书馆
The Library

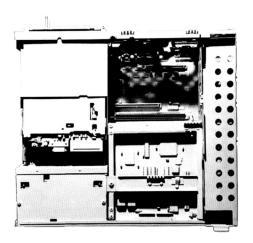

图书馆是在老馆北侧建的一个新馆。新馆内设有书目大厅，阅览室和古善本收藏与展示空间，馆藏图书800万册，是一个具有信息化时代特征和功能的现代图书馆。图书馆的整体空间组合和布局原初于计算机，其富有当代技术特征的内部构造关系可以作为图书馆的文化表达的内涵。

The library is a new building newly built on the north side of the old library. Inside the new library there are catalogue hall, reading room, collection and display room for old and rare books. With 8 million books, it is a modern library with time feature and function of the information age. The overall space combination and layout originates from computers, whose internal structural relation full of contemporary technology features can serve as content for cultural expression of libraries.

图书馆鸟瞰图
Aerial view of the Library

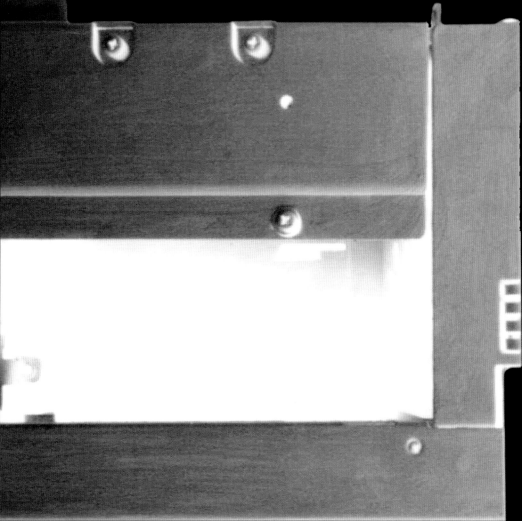

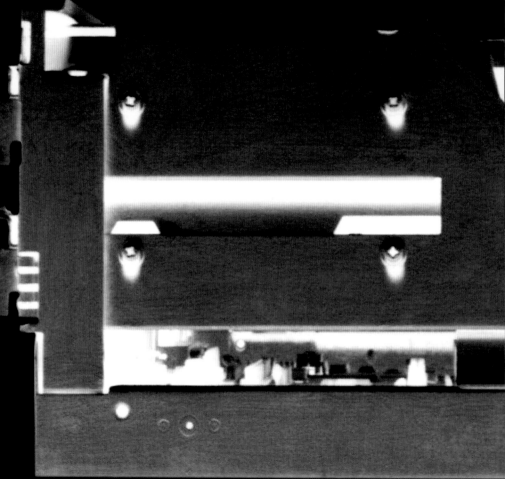

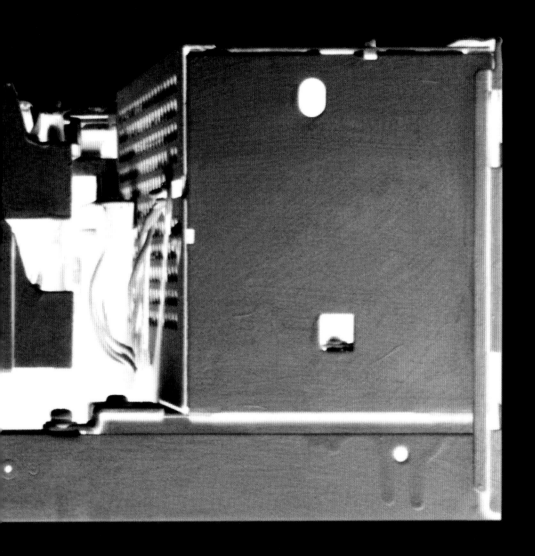

图书馆东立面
East facade of the Library

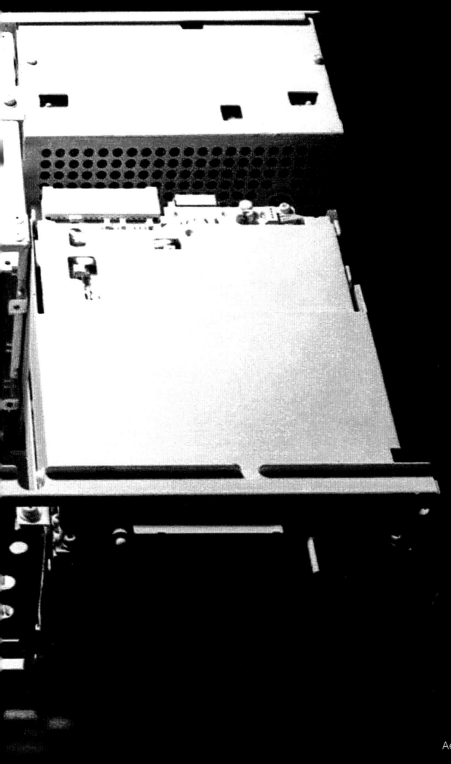

图书馆鸟瞰图
Aerial view of the Library

图书馆西侧立面
West facade of the Library

图书馆鸟瞰图
Aerial View of the Library

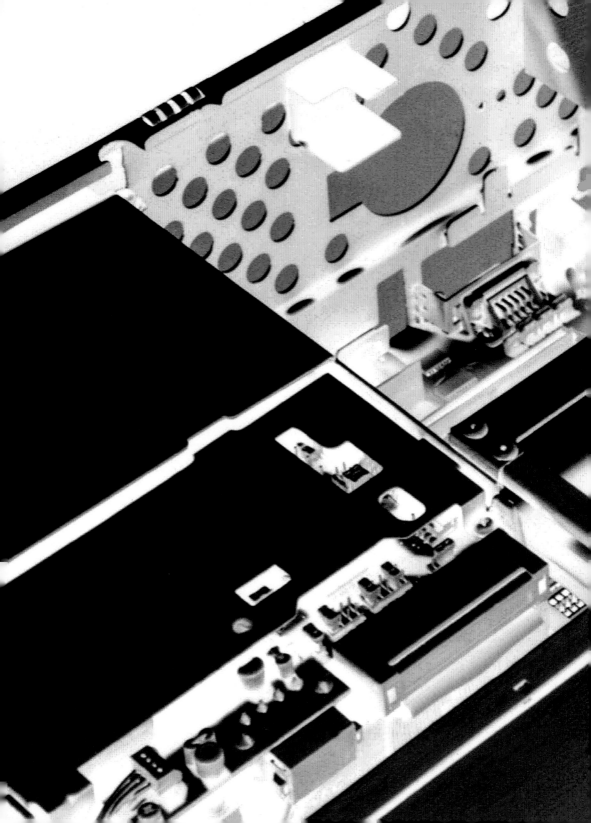

图书馆鸟瞰图
Aerial view of the Library

图书馆鸟瞰图
Aerial view of the Library

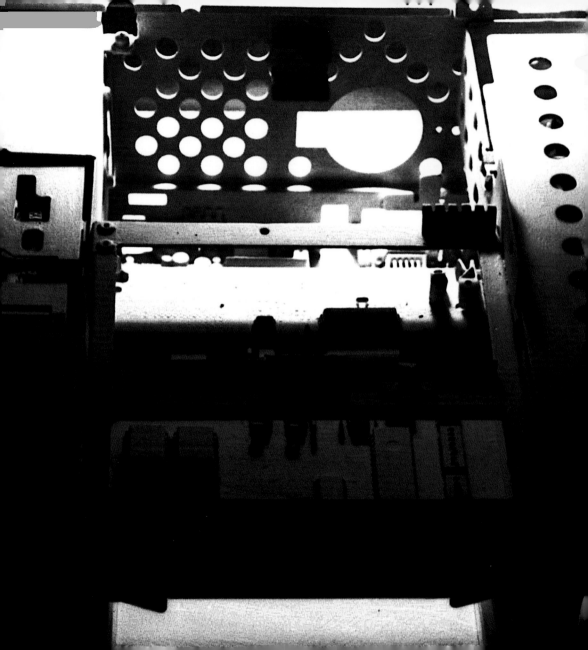

图书馆鸟瞰图
Aerial view of the Library

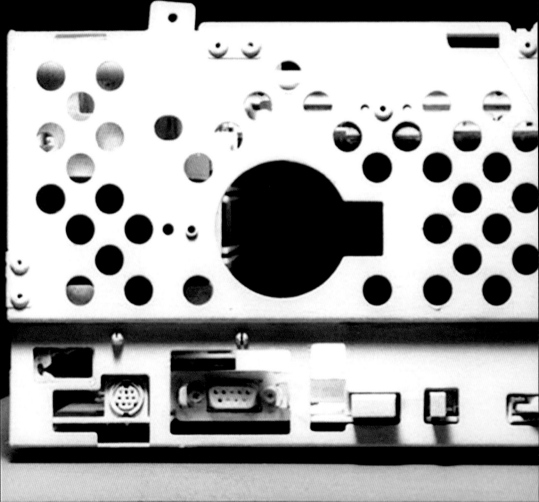

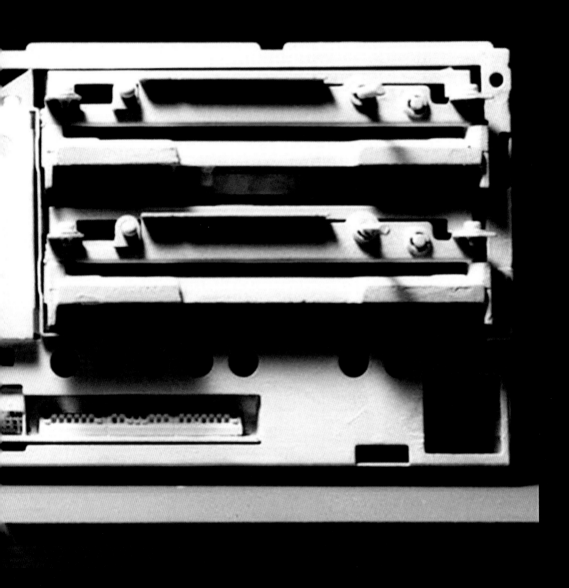

图书馆南立面
South facade of the Library

图书馆西北角中庭透视图
Perspective drawing of atrium at northwest corner of the Library

白天的图书馆南侧中庭采光窗
Atrium lighting window on the south of the Library in the day

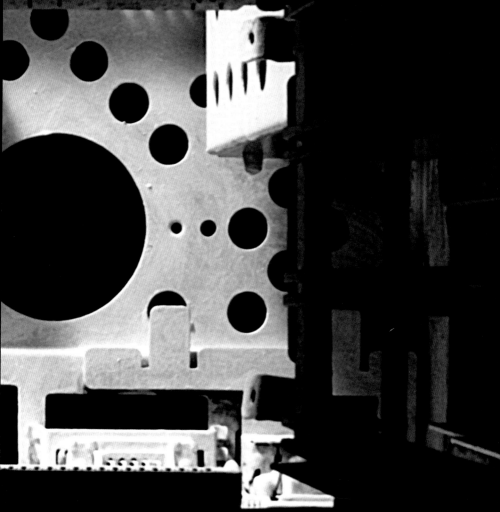

图书馆藏书库
Book warehouse of the Library

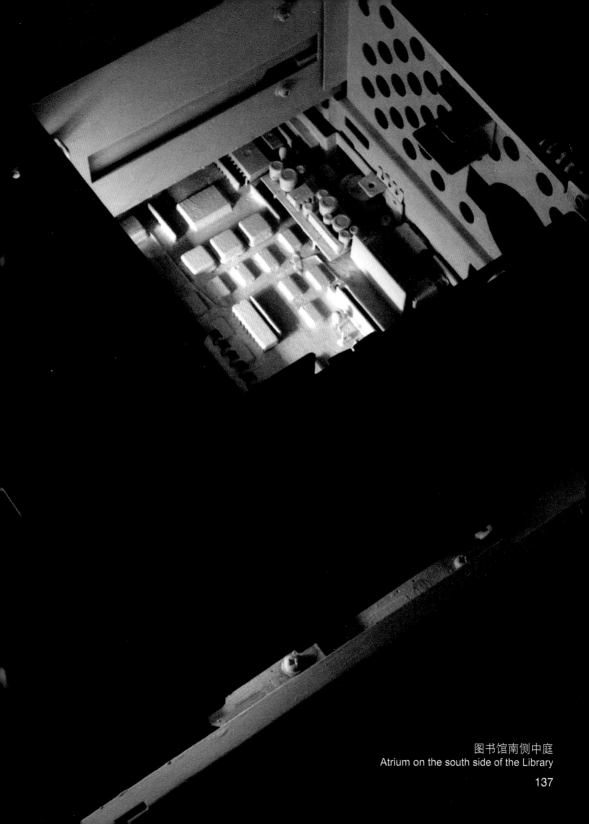

图书馆南侧中庭
Atrium on the south side of the Library

图书馆南侧中庭内开敞空间
Open space inside atrium on the south side of the Library

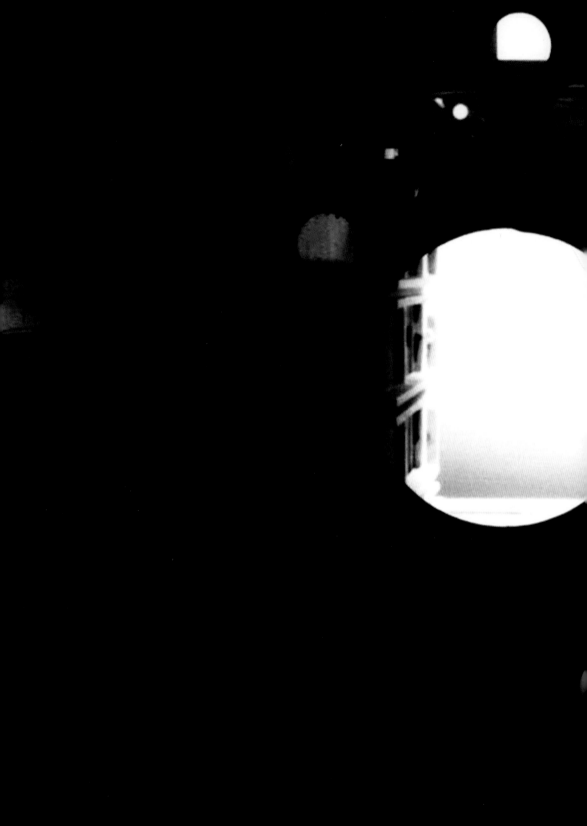

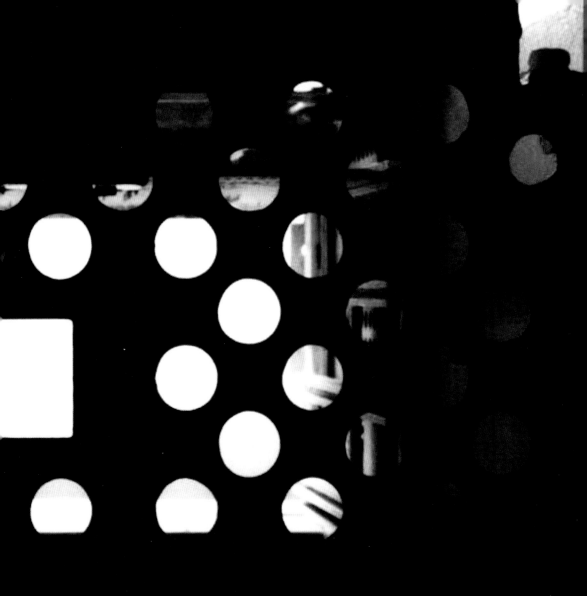

从南侧室外透过采光窗看向图书馆内部
Internal view by looking through lighting window from outdoors on the south side of the Library

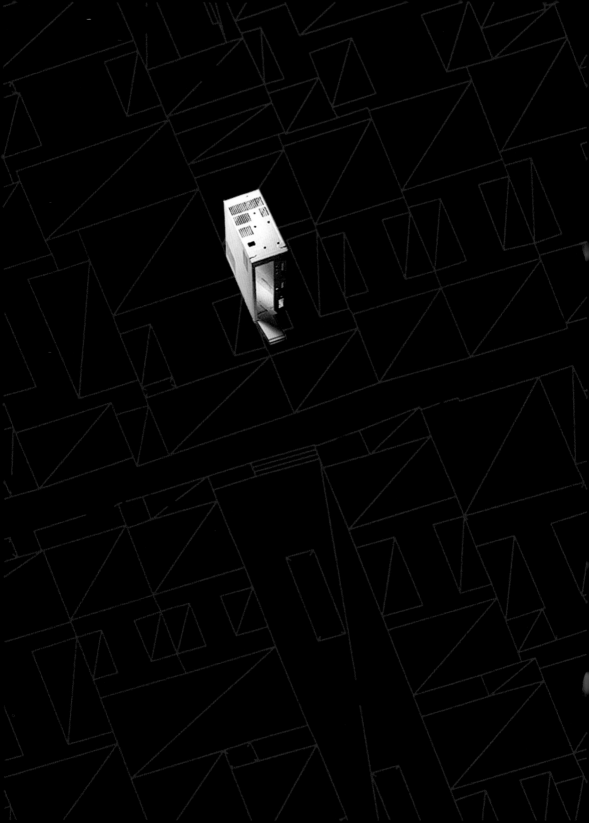

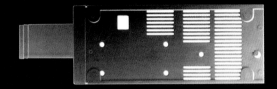

光宅
House of Light

光宅是对旧城中老旧住宅空间的改造案例，改造不仅仅是一个自上而下组织起来的实施项目，而是将居住者组织和行动起来的一个自下而上的实施项目。

House of Light is a case of renovating old houses in old city. The renovation is a project not only organized from top to bottom but also organizing and involving residents from bottom to top.

白天的光宅

House of Light in the day

夜晚的光宅
House of Light at night

光宅透视
Perspective drawing of House of Light

白天光宅作为光的反射装置
House of Light in the day as light reflection set

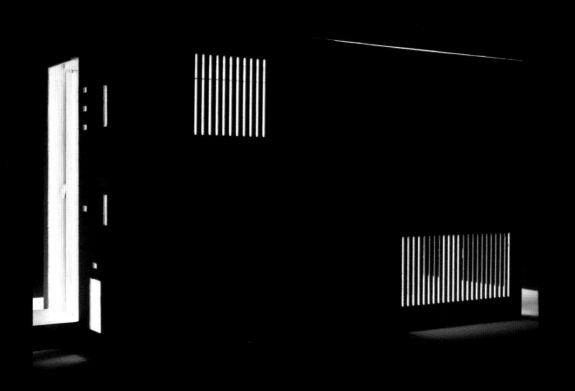

夜间光宅成为光的发射装置
House of Light at night as light transmitting set

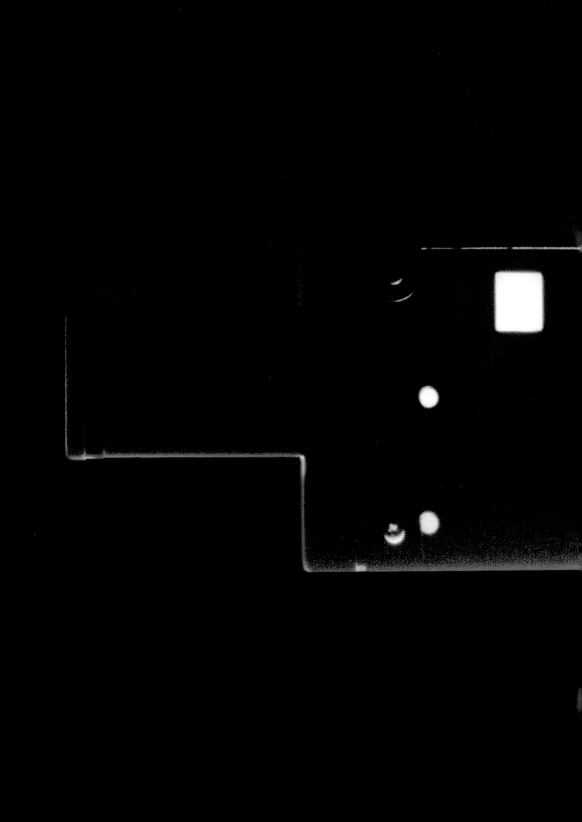

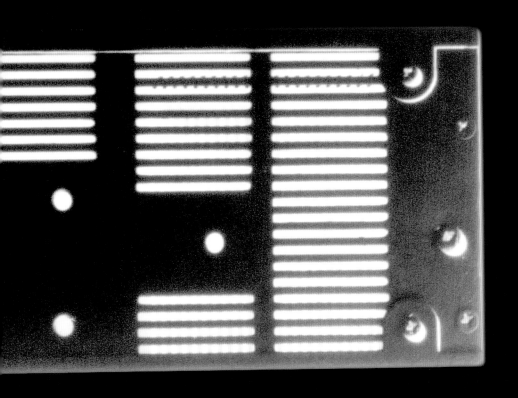

光宅屋顶平面
Roof plan of House of Light

光通过光孔射出光宅
Light goes out from House of Light through a hole

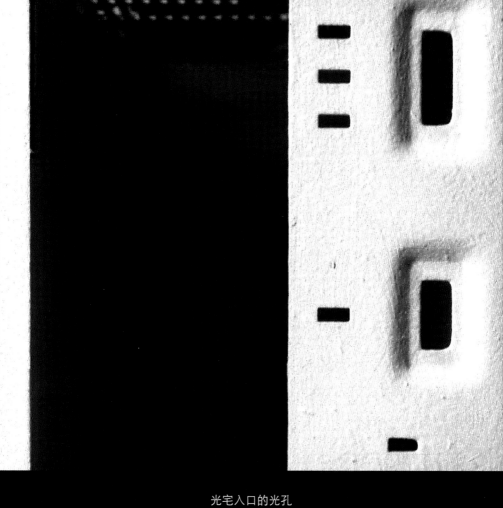

光宅入口的光孔
Light hole at the entrance of House of Light

10

文化宫
Cultural Palace

文化宫
Cultural Palace

文化宫位于复兴门立交桥西北侧,紧邻北京二环路城市快速路,是以电路板的构成关系作为文化宫的形态表征。文化宫整体为南北方向布局,是集艺术、科技、文化、信息功能为一体的综合体。
Cultural Palace is at the northwest side of Fuxingmen Overpass, close to the Second Ring Road city expressway of Beijing. It uses the structure of circuit board as its pattern feature. With an overall south-north layout, Cultural Palace is a comprehensive site integrating functions of art, science, culture and information.

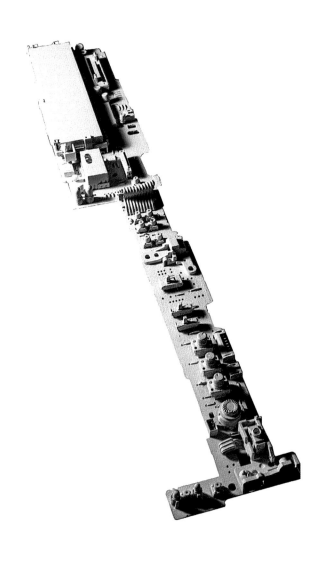

文化宫以一个漂浮的特征坐落在城市中
Cultural Palace is like a floating symbol in the city

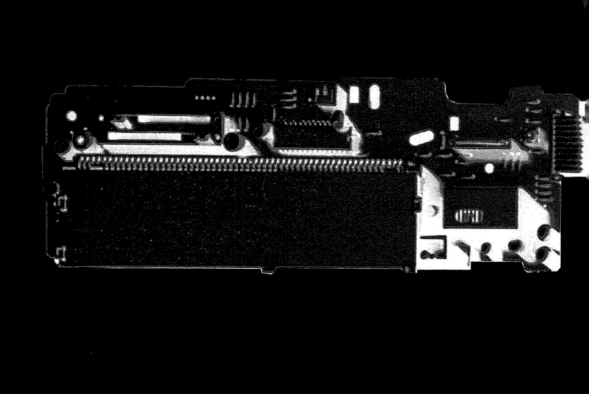

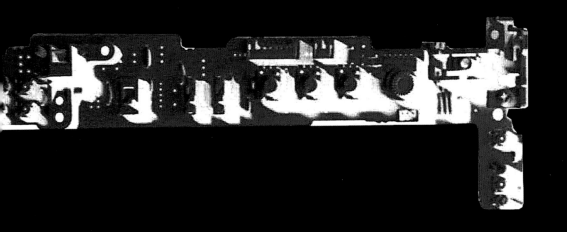

文化宫的设计意向
Design intention of Cultural Palace

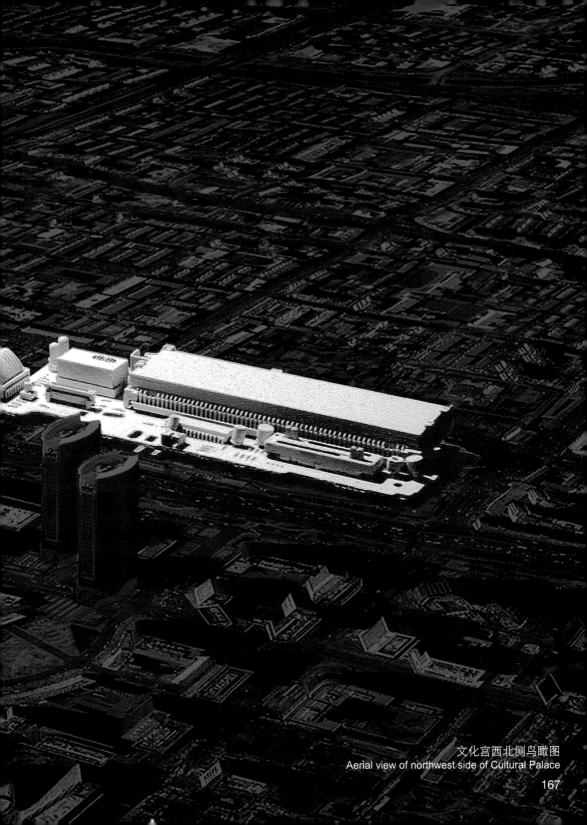

文化宫西北侧鸟瞰图
Aerial view of northwest side of Cultural Palace

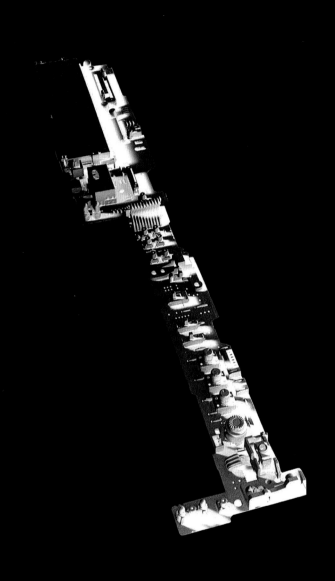

文化宫的设计意向
Design intention of Cultural Palace

11
大木仓密集城市
Damucang Intensive City

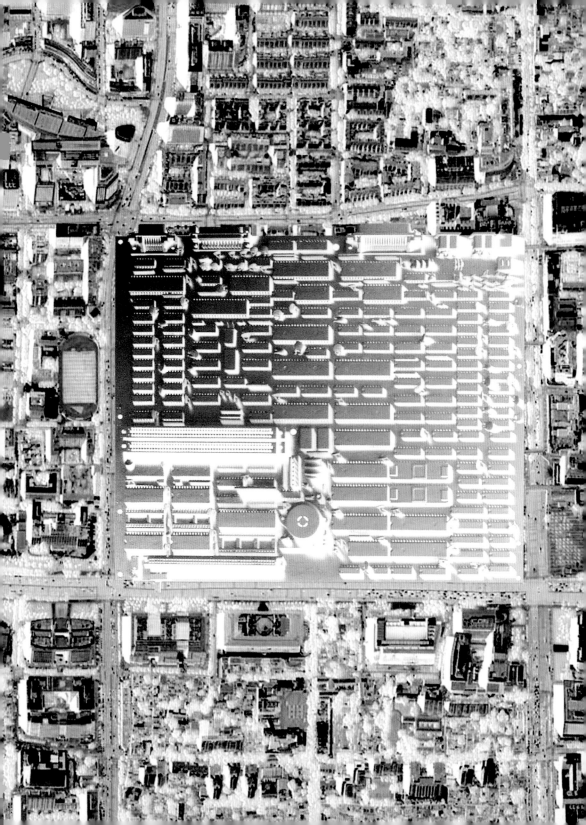

大木仓密集城市
Damucang Intensive City

大木仓地区是北京西城胡同聚集的区域。随着城市的发展，新的规划同样以低矮的方式保持原有社区的特征，这是方案思考的切入点。整体的形态原初于计算机的电路板，计算机内部的合理性布局同样可以与社会的合理关系互为映照。

Damucang area in Xicheng of Beijing is densely crisscrossed by many alleys. With urban development, new design still uses low buildings to maintain the original community feature, which is the entry point of the plan consideration. The overall form originates from circuit boards of computers, as the scientific layout inside computers and social rational relation can cast light on each other.

大木仓密集城市鸟瞰
Aerial view of Damucang Intensive City

大木仓密集城市鸟瞰
Aerial view of Damucang Intensive City

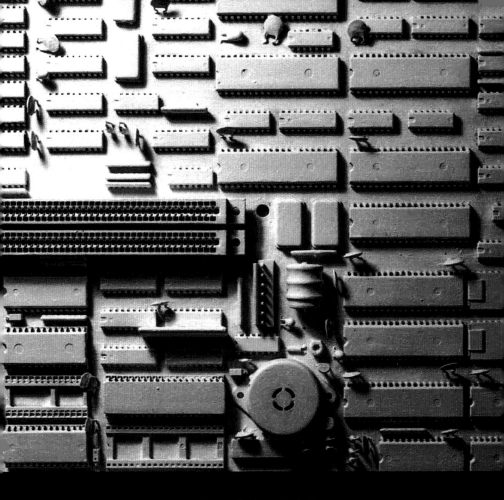

大木仓密集城市局部鸟瞰
Aerial view of partial Damucang Intensive City

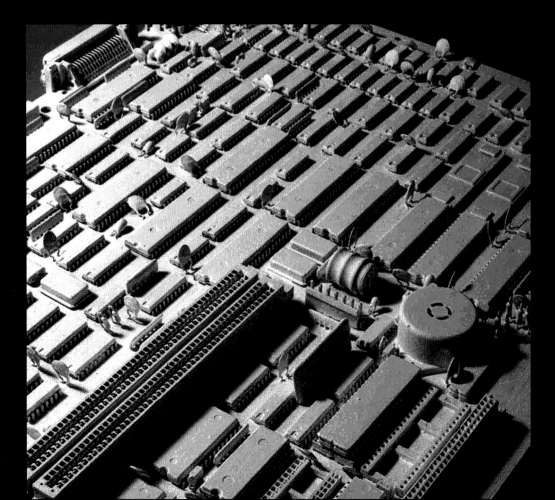

希望小学
Hope Primary School

希望小学位于城市的社区中心,采用芯片插座为学校的原初形态。
Hope Primary School is located at the city community center, using chip carrier socket as the original pattern.

希望小学鸟瞰
Aerial view of Hope Primary School

希望小学鸟瞰
Aerial view of Hope Primary School

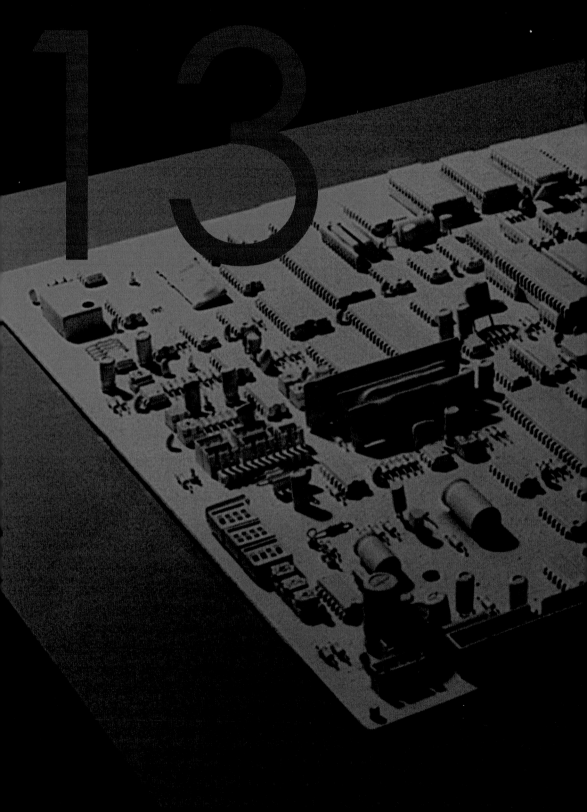

大观园规划
Grand View Garden Plan

大观园规划位于通州新城中心地区，是一个具有未来特征的综合性的生活与商业一体化社区。
Grand View Garden Plan, located at the center of Tongzhou New City, is a comprehensive community integrating life and business with future features.

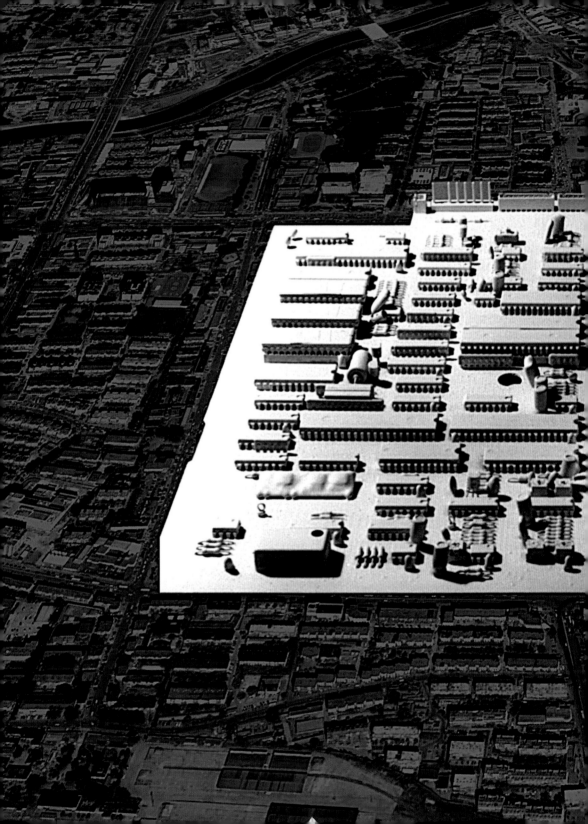

大观园总体鸟瞰
Overall aerial view of Grand View Garden

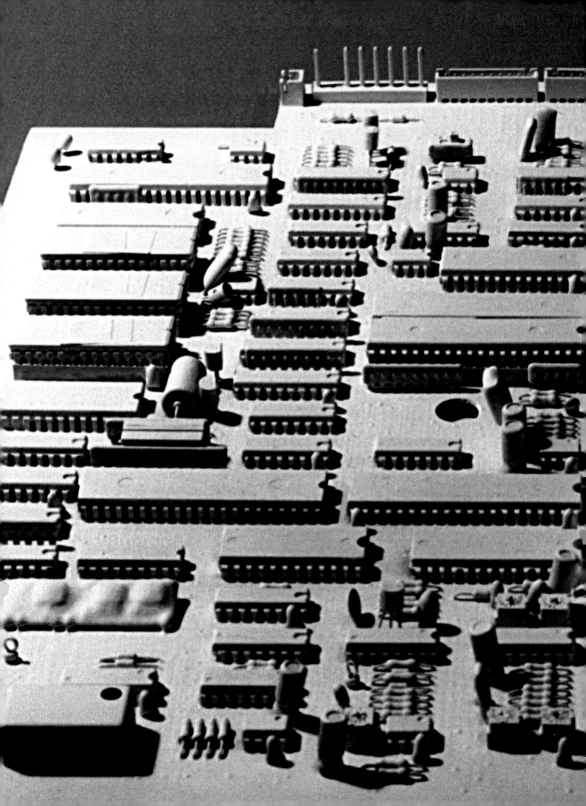

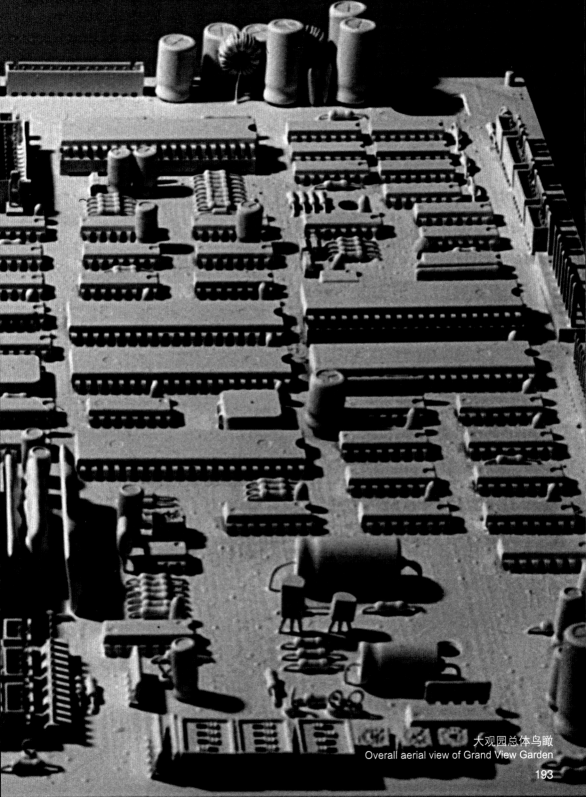

大观园总体鸟瞰
Overall aerial view of Grand View Garden

数学家住宅
Mathematicians Residence

数学家住宅
Mathematicians Residence

该住宅是一个起居室面向侧向庭院的小建筑，室内由卧室、书房和一个小型卡拉ok厅所构成。
This is a small building consists of bedroom, reading room and a small Karaoke room, with its sitting room facing a courtyard on side.

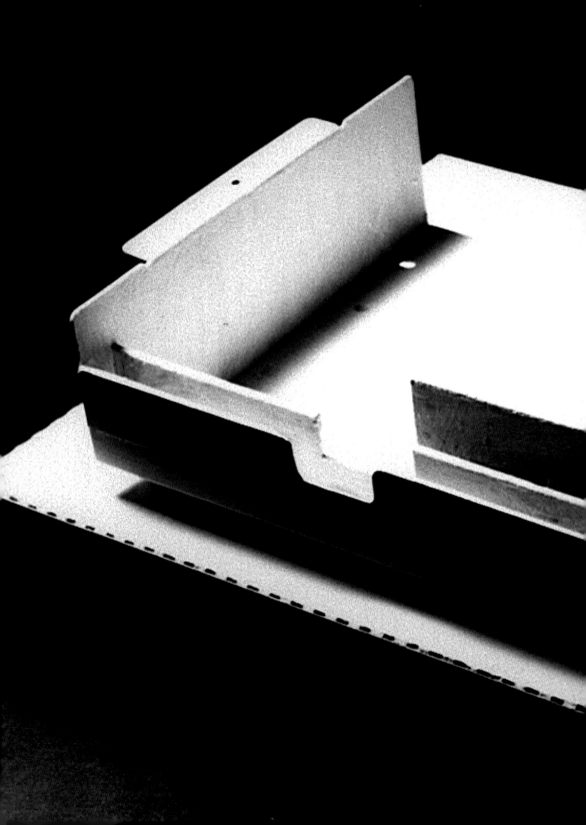

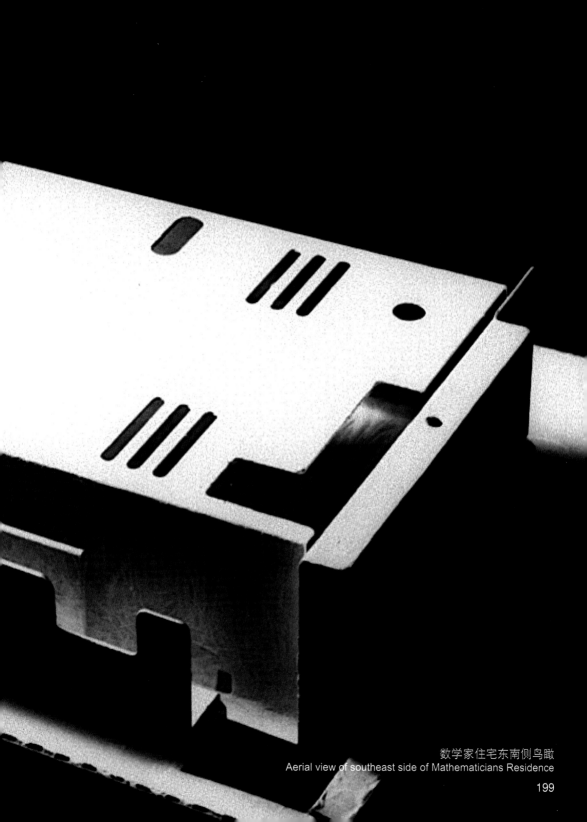

数学家住宅东南侧鸟瞰
Aerial view of southeast side of Mathematicians Residence

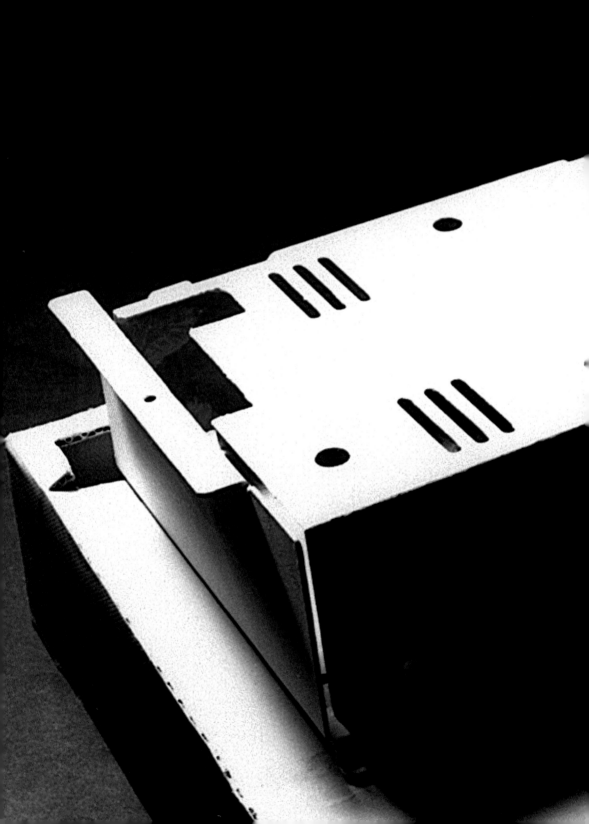

数学家住宅东北侧鸟瞰
Aerial view of northeast side of Mathematicians Residence

数学家住宅鸟瞰
Aerial view of Mathematicians Residence

贸易市场
Trade Market

这是为北京新发地市场拟新建的综合性农副产品贸易市场所做的方案设计。该市场占地面积3万平方米，设计采用一个整体的大空间布局，北侧一端设三个办公用房，大空间的上部是一个绿色的屋顶花园。

This is a plan designed for a comprehensive agricultural and sideline products trade market to be built by Beijing Xinfadi Market. This market, occupying a land area of 30,000 square meters, is designed to have one integral big space, with three office sites on its north and one green roof garden on top.

贸易市场鸟瞰
Aerial view of Trade Market

贸易市场鸟瞰
Aerial view of Trade Market

贸易市场屋顶花园
Roof garden of Trade Market

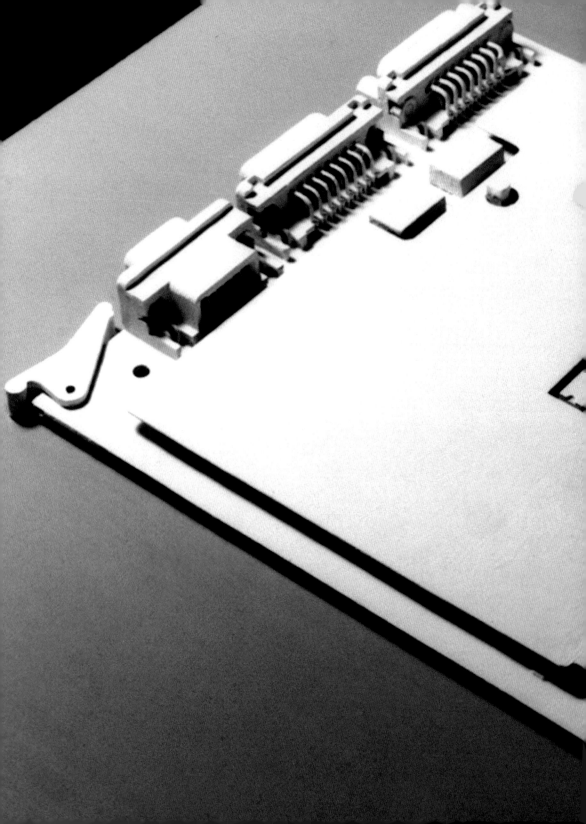

贸易市场鸟瞰
Aerial view of Trade Market

16

公园住宅
Park Residence

石景山公园住宅位于道路一侧的斜面坡地，所有住宅单元即可远眺公园风景，又可确保邻居之间的隐私不被干扰。住宅的下层有一个公共的交流空间，安放有社区活动中心、咖啡式图书中心及便利店。

Shijingshan Park Residence is located on a slope at one side of a road. From every residence unit one may view park scenes at a distance without disturbing neighbors' privacy. At the lower level of the building there is a public space for exchange and communication, equipped with community event center, café-style reading center and

公园住宅鸟瞰图
Aerial view of Park Residence

公园住宅鸟瞰图
Aerial view of Park Residence

公园住宅鸟瞰图
Aerial view of Park Residence

17

摄影家沙龙
Photographers Saloon

摄影家沙龙
Photographers Saloon

摄影家沙龙是设立在北京车公庄大街路南侧的一个小型交流场所，建筑内部设有对外展览空间，同时建筑内部还设有面向社会的摄影教学的教室。建筑的整体充满了镜头光圈一般的圆形窗口，形成了一个充满了光线的感光盒子。

Photographers saloon is a small site for exchange on the south side of Beijing Chegongzhuang Avenue. There are exhibition rooms and teaching rooms for photography inside the building. The overall character resembles a photosensitive box full of light rays, with round windows similar to lens aperture.

摄影家沙龙鸟瞰图
Aerial view of Photographers Saloon

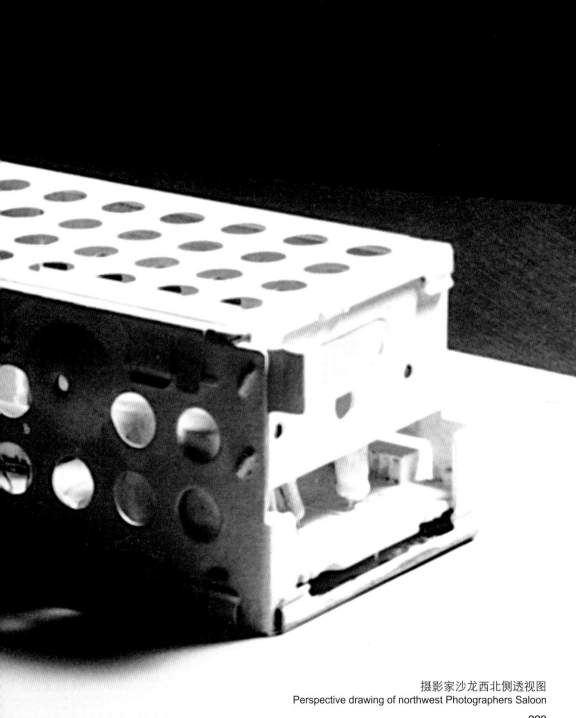

摄影家沙龙西北侧透视图
Perspective drawing of northwest Photographers Saloon

18
未来城
Future City

未来城鸟瞰
Aerial view of Future City

未来城位于北京的中轴线最北端，位于奥林匹克公园以北的延长线上。未来城本体是一个巨构的综合体，也是一个立体城市。该巨构建筑总高为900米，综合体拥有空中交通连线，最顶端是一个直升机的停机坪，使得空中的航线可以方便地与综合体本身取得直接的联系。此外，停机坪下面是一个开敞的空中庭院，建筑间采用新发明的软性交通相联结。未来城内设有办公、住宅、电影院、商业，建筑的表皮采用可以作为太阳能集热的材料系统。未来城预计于2050年完成，届时将容纳90万人在这里生活和居住。

Future City is located on the northernmost side of central axis of Beijing, the northward extension line from Olympic Park. Future City itself is a giant complexity and three-dimensional city. This giant construction, with a total height of 900 meters, possesses its own air traffic line. At its very top there is a parking apron for helicopters, enabling direct access between air traffic line and the complexity itself. Under the parking apron there is an open midair courtyard, with building materials interconnected by newly-invented soft transportation. Inside Future City there are office buildings, residential buildings, cinema and business facilities, with architecture facade using solar-energy generating material. Future City is expected to be completed by 2050, hosting 900,000 people to live here.

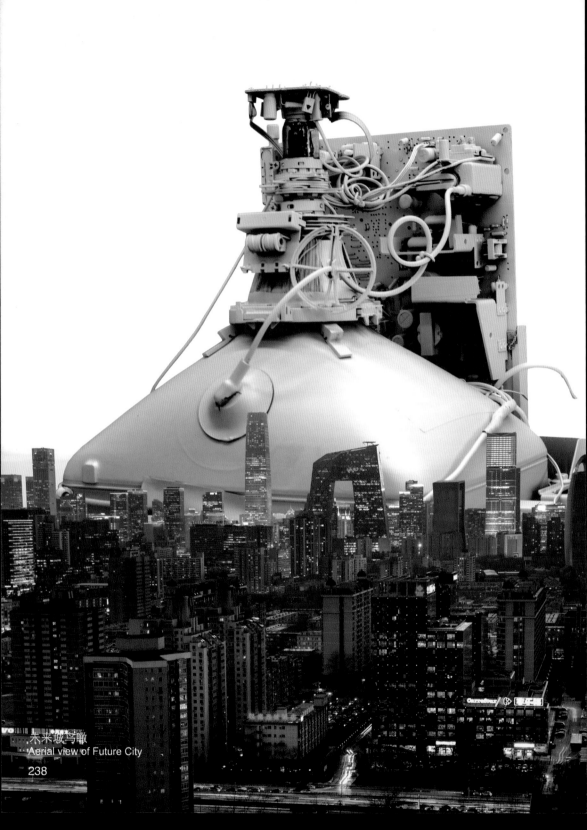

未来城鸟瞰
Aerial view of Future City

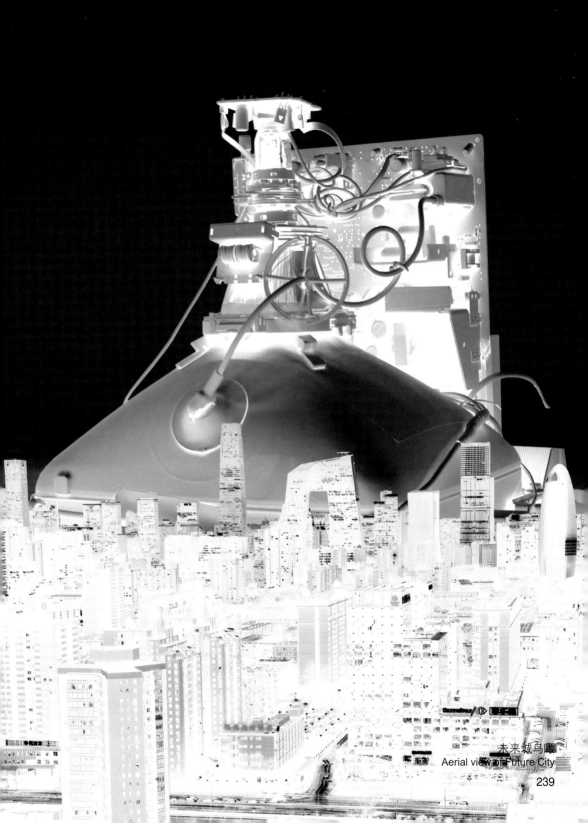

未来城鸟瞰
Aerial view of Future City

未来城鸟瞰
Aerial view of Future City

未来城北立面
North facade of Future City

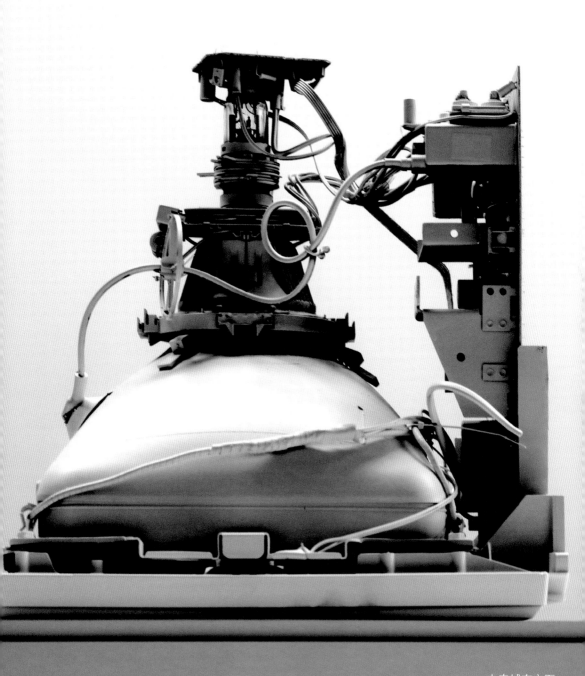

未来城东立面
East facade of Future City

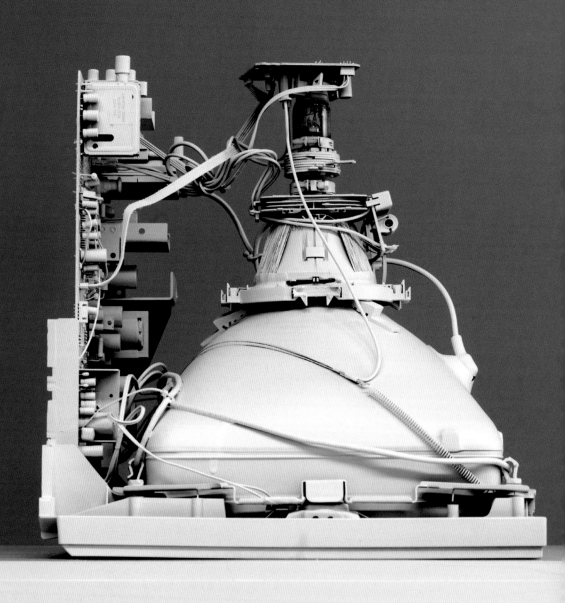

未来城西立面
West facade of Future City

未来城东北角
Northeast corner of Future City

"废物"观察术　　　　　　　　　　　　　　　How to Observe "Junk"

左图：现代美术馆设想
Left figure: Concept of Modern Art Gallery

    前面18个由被称为"废物""垃圾"的结晶体所构成的未建成的建筑实例，呈现给我们的是一个个富于时代感的"场景"。尽管这些建筑在传统建筑教养的眼光下很可能是"丑陋"的，但这一系列场景突兀地呈现在我们眼前，一种时代记忆的"唤起"似已开始在大脑中形成。这18个场景来源于对一个个"垃圾"本身的尺度进行收放时，而获得的新的在建筑层面上的意义。

    其实人的眼睛所能够看到的世界范围和形态实际上在不断地发展。比如由于镜头的产生，门采尔绘画中关于对象物的描写就会进一步变得具象和深入。伦勃朗对光线的运用似乎更像一个不好的镜头（四个角的光晕很浓的镜头）后面所看到的场景。工具其实对于人类的世界的认识变得非常重要。由于"进步"，人类生产出的工具，产品丰富并有时代特征，特别是工业产品，本身的形态就是一种新的形态。

    在这样的一种前提下，时代的造型特征丰富多变。比如20世纪开始的大工业的迅猛发展，产品伴随着其功能的变化不断地更新改造，于是乎一种

The above mentioned unestablished architectural cases, constituted by crystallized objects of "junk" and "rubbish", provide us one by one with "scenes" full of sense of the times. Although they may be considered "ugly" in conventional architectural sense, however, when this series of scenes appear abruptly in front us, an "evocation" of memory of certain ages seems to take shape in mind. These 18 scenes offer new architectural meanings, which are obtained from changing the scale of "rubbish" objects.

Actually the scope and pattern of the world that human eyes can perceive are developing continuously. For example, with the invention of lens, the description on objects in paintings by Menzel is becoming more concrete and in-depth. The utilization of light rays by Rembrandt is like the scene behind a poor lens whose corners are dotted by thick halos. Tools actually become extremely important for mankind to understand the world. Due to "progress", tools produced by mankind are rich in variety and distinctive in feature of times. Particularly, for industrial products, their patterns are, on their own, a new type of pattern.

With such a premise, the pattern and feature of ages are rich and variable. For example, the rapid development of large-scale industry in the 20th century leads to continuous transformation and renovation of products along with functional changes, therefore a so-called pattern of time feature quickly develops,

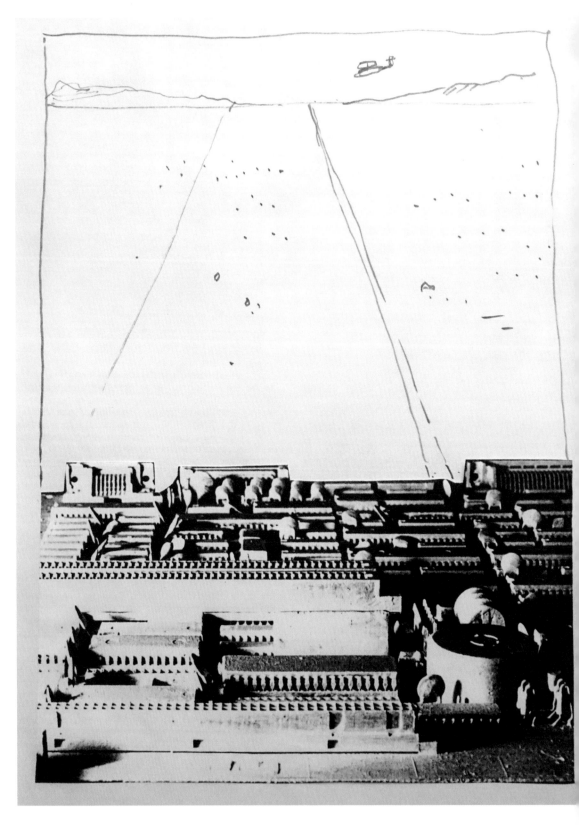

左图：荒漠中的新城设想
Left figure: Concept of new city in wilderness

所谓的时代特征的造型便迅速地成为"结晶体"的卖点而迅速地发展。而这样的发展的结果，客观上产生了一大批被用过的"结晶体"的形态。而作为建筑本身，由于其产生的时间特点，往往落后于迅速变化着的"结晶体"的形态，因此，一系列"结晶体"的形态本身便成为代替自然的新的形态，而原有的"结晶体"便成为了"废物"和"垃圾"，而当我们对"废物"和"垃圾"进行肢解和观察的时候，呈现在我们眼前的这些垃圾，往往拥有着复杂的构造，而形成这些对象物的所有"零件"彼此之间有着紧密而严谨的关联性，在夺人眼的同时，还诉诸着规则之美，同时整体又极具有时代特征。

我们必须要说的是对于观察尽管已经成为了垃圾的"结晶体"其实必须要拥有一种"开放"的心态。这种"开放"的心态应该抛开以往的由"内部向外"的建筑思考，有时在对已经成为了垃圾的"结晶体"的读解的过程中，捕获其呈现并诉诸的新的含义。

确切地说，对于眼前的"垃圾"如何看和如何选是重要的。如果要使"垃圾"与现实的建筑发生关联或成为现实，一系列的"观念"与"文化"的"抛

using "crystallized object" as its selling point. As a result of such a development, a large number of used "crystallized objects" come into existence. Architectural projects, due to the time of establishment, often fall behind the fast-changing "crystallized objects" pattern. Therefore, a series of "crystallized objects" themselves become new patterns in the place of nature, while the former "crystallized objects" become "junk" and "rubbish". When we disassemble and observe these "junk" and "rubbish", we find them possess complicated structure, close and strict interrelation among "components", which boasts great attraction, beauty of rules and overall strong feature of ages.

It must be pointed out that when observing "crystallized objects" that already became rubbish, we must hold an "open" mindset, getting rid of the former way of architectural thinking "from internal to external". Sometimes during interpreting "crystallized objects" that became rubbish, we need to capture new meanings they present and convey.

To be precise, it is important how to regard and choose "rubbish" in front of our eyes. If we want to realize the connection between "rubbish" and architecture in reality, it is important to have a series of procedures of "discarding" actions of "concepts" and "culture". Only after a series of discarding procedures can new scene emerge. "Replacement" enables new scale to match new pattern, while new "regeneration" will definitely

NEC 9801 計算機の解体問モ　97年1月

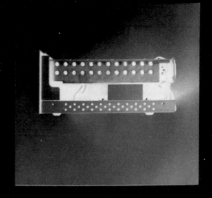
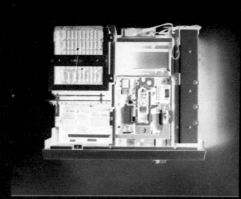
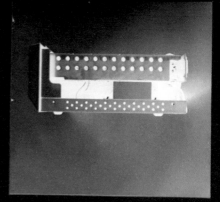
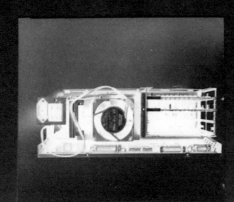
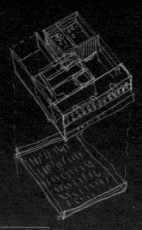

左图：1997年1月NEC 9801计算机肢解现场
Left figure: Disassembling NEC 9801 computer in January 1997

右图：照相机的剖面置换为住宅的剖面
Right figure: Section of a camera transformed into section of a residence

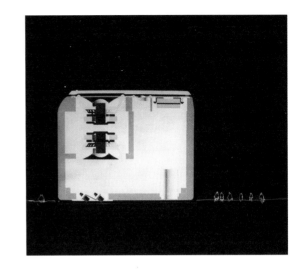

掉"过程是重要的。在经过一系列的抛弃过程之后，新的风景才能够呈现，"置换"使得新的尺度对应新的形态，而新的"重生"也一定会成为新的传统。

记得儿时的最大缺点就是经常被父母批评"不爱惜东西"，因为所有的玩具会在玩过一阵子之后便对其进行肢解，似乎总希望看到表皮之后的场景和世界。而表皮背后的世界又常常的感觉新于表皮本身已经被看够的厌倦的场景。

"新"的再更新，新亦瞬间变幻，而对于旧物的拆解与观察或许正在展示着一个宝库般的视觉"矿藏"。

本书的这18个案例就是经过一系列拆解与观察的产物：现代美术馆设计是曾经陪伴我完成了外语学习的"磁带播放机"，之后尽管其成为了废物垃圾，但却展示出建筑层面上的空间魅力。而这种特有的空间关系超越了其本身原有的仅仅作为播放声音的功能性。案例中的当代博物馆，实际上是机上读取磁盘的装置，将其打开拉伸，博物馆的形态就呈现在那里。新世纪媒体中心是电脑上的显示器，将其倒置的瞬间，"如果能够在北京的二环有这样的

become new tradition.

I remember I was criticized often by my parents in childhood for my biggest shortcoming as "not taking good care of stuff", because I almost disassembled all my toys after playing a while, expecting to see scenes and worlds behind their shells. The world behind shell often presented fresher and newer feelings than scenes on shell that I got tired of.

Renew the "new", the new also changes instantly. By disassembling and observing old objects, we may find a treasure-like visual "mineral resources".

The 18 cases of this book are results after a series of disassemble and observation: the design of Modern Art Gallery is based on a cassette player that accompanied my foreign language study, which later became a junk. However, it displays charm of space in terms of architectural sense. Its unique spatial relation transcended its original functionality of purely playing audio. Contemporary Museum is actually based on an equipment to read discs, which, being opened and extended, displays the pattern of the museum. New Century Media Center is based on a display device of a computer which, at the moment of being placed upside down, inspired me in a flash with a thought that "it would be very charming to have such a building on the second ring road of Beijing!" Science Mansion is based on a tool to read big disc in a computer,

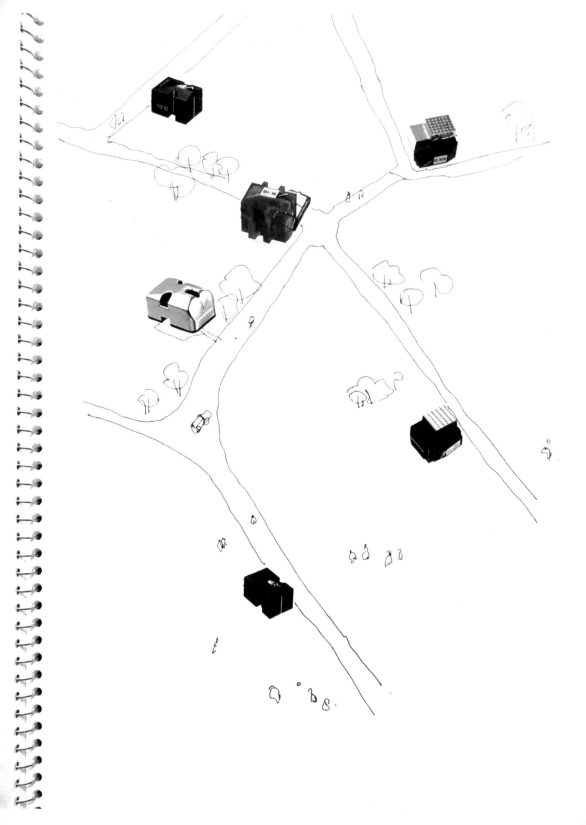

左图：6个小住宅的同时设计
Left figure: Concurrent design of 6 small residences

一个建筑将会是多么的有魅力"的念头瞬间闪过。科技大厦是计算机中读取大磁片的道具，其暴露的元器件的组合关系呈现着严密的科技逻辑。将其坐落在西单广场本身就会拥有科技的教意。CBD国际城的规划形态本身是一台传真机的线路底板，本身就已经拥有着收发和传递多种信息的寓意。从中不难发现广场、商业中心、美术馆等城市的组合要素，疏密的结合提示着未来的城市的走向。都会广场是计算机内部摆放磁盘的搁架，内部蕴藏丰富的特征，与商业的性格直接吻合。光宅本身就是包裹磁盘外壳的本身，寓意着住宅是居住的机器。曾经包裹过读取地盘机器的表皮躯体同样地可以包裹人类这一个肉体的机器。国家大剧院是录像机的机芯，复杂的空间关系，播放影像的功能形态所给予的结晶体性格本身，作为大剧院当之无愧……

  如果这个世界还在探讨所谓具有时代性造型的问题的话；如果追求造型而不究取造型原委的话；如果连牛粪（当然是造型美的牛粪）都可以成为建筑摹本的话，周围的一切（包括垃圾）也都将可以成为建筑造型的摹本。

whose combination of exposed components reveals very strict logic of science and technology. Such a building at Xidan Square will possess strong sense of science and technology. CBD International City is based on the circuit board of a fax machine that contains the meaning of receiving and transmitting multiple information, from which we can discover not only urban elements such as square, business center, art gallery but also the future city development trend indicated by the combination of dense and sparse arrangement. Metropolis plaza is based on a tray inside a computer to hold discs, which contains rich characters directly in line with business characters. House of Light itself, based on the shell around discs, implies that residence is the machine for living. The shell once enveloped disc-reading machine can be transformed into a machine enveloping human flesh. National Grand Theatre is based on the mechanism of a video recorder with complicated spatial relation. The character of the crystallized objects endowed by the function pattern of video playing is highly worthy of grand theatre.

Everything around us (including rubbish) can become prototype of architectural image, if this world is still talking about so-called images with features of times, if such pursuit of images is free from tracing the origin of such images, if even cow dung (of course, cow dung with beautiful modeling) can become architectural prototype.

作者简介 About the author

王昀简介

王昀 博士
1985年　毕业于北京建筑工程学院建筑系
　　　　获学士学位
1995年　毕业于日本东京大学
　　　　获工学硕士学位
1999年　毕业于日本东京大学
　　　　获工学博士学位
2001年　执教于北京大学
2002年　成立方体空间工作室
2013年　创立北京建筑大学建筑设计艺术研究
　　　　中心担任主任
2015年　于清华大学建筑学院担任设计导师

建筑设计竞赛获奖经历：
1993年　日本《新建筑》第20回日新工业建筑设计竞赛获二等奖
1994年　日本《新建筑》第4回S×L建筑设计竞赛获一等奖

主要建筑作品：
善美办公楼门厅增建，60m² 极小城市，石景山财政局培训中心，庐师山庄，百子湾中学，百子湾幼儿园，杭州西溪湿地艺术村H地块会所等

参加展览：
2004年　"'状态'中国青年建筑师8人展"
2004年　首届中国国际建筑艺术双年展
2006年　第二届中国国际建筑艺术双年展
2009年　比利时布鲁塞尔"'心造'——中国当代建筑前沿展"
2010年　威尼斯建筑艺术双年展德国卡尔斯鲁厄Chinese Regional Architectural Creation建筑展
2011年　捷克布拉格中国当代建筑展，意大利罗马"向东方——中国建筑景观"展，中国深圳·香港城市建筑双城双年展
2012年　第十三届威尼斯国际建筑艺术双年展中国馆

WangYunProfile

Dr. Wang Yun
Graduated with a Bachelor's degree from the Department of Architecture at the Beijing Institute of Architectural Engineering in 1985.
Received his Master's degree in Engineering Science from Tokyo University in 1995.
Received a Ph.D. from Tokyo University in 1999.
Taught at Peking University since 2001.
Founded the Atelier Fronti (www.fronti.cn) in 2002.
Established Graduate School of Architecture Design and Art of Beijing University of Civil Engineering and Architecture in 2013, served as dean.
Served as a design Instructor at School of Architecture, Tsinghua University in 2015.

Prize:
Received the second place prize in the "New Architecture" category at Japan's 20th annual International Architectural Design Competition in 1993
Awarded the first prize in the "New Architecture" category at Japan's 4th SxL International Architectural Design Competition in 1994

Prominent works:
ShanMei Office Building Foyer, 60m2 Mini City, the Shijingshan Bureau of Finance Training Center, Lushi Mountain Villa, Baiziwan Middle School, Baiziwan Kindergarten, and Block H of the Hangzhou Xixi Wetland Art Village.

Exhibitions:
The 2004 Chinese National Young Architects 8 Man Exhibition, the First China International Architecture Biennale, the Second China International Architecture Biennale in 2006, the "Heart-Made: Cutting-Edge of Chinese Contemporary Architecture" exhibit in Brussels in 2009, the 2010 Architectural Venice Biennale, the Karlsruhe Chinese Regional Architectural Creation exhibition in Germany, the Chinese Contemporary Architecture Exhibition in Prague in 2011, the "Towards the East: Chinese Landscape Architecture" exhibition in Rome, the Hong Kong-Shenzhen Twin Cities Urban Planning Biennale, Pavilion of China The 13th international Architecture Exhibition la Biennale di Venezia in 2012.

www.fronti.cn